Part Programming Techniques

Graham T. Smith

CNC Machining Technology

Volume III
Part Programming Techniques

With 36 Figures

Springer-Verlag
London Berlin Heidelberg New York
Paris Tokyo Hong Kong
Barcelona Budapest

Graham T. Smith
Technology Research Centre, Southampton Institute,
City Campus, East Park Terrace, Southampton SO9 4WW,
UK

Cover illustration: Ch.1, Fig.34d. Tool path graphics – illustrating cutter motions superimposed on a "wire-frame model". [Courtesy of McDonnell Douglas/Boston Digital Corporation.]

ISBN 3-540-19830-X Springer-Verlag Berlin Heidelberg New York
ISBN 0-387-19830-X Springer-Verlag New York Berlin Heidelberg

British Library Cataloguing in Publication Data
A catalogue record for this book is available from the British Library

Library of Congress Cataloging-in-Publication Data
A catalog record for this book is available from the Library of Congress

© Springer-Verlag London Limited 1993
Printed in Germany

Typeset by Best-set Typesetter Ltd., Hong Kong
69/3830-543210 Printed on acid-free paper

*To my grandfather
Mr T.W. Chandler
who encouraged me
to take an interest
in all things*

ΣΟΦΟΣ ΑΝΗΡ Ο ΕΞ ΙΔΙΑΣ
ΠΕΙΡΑΣ ΔΙΔΑΣΚΟΜΕΝΟΣ
ΣΟΦΟΤΑΤΟΣ ΔΕ Ο ΕΚ
ΤΗΣ ΤΩΝ ΑΛΛΩΝ

Translation:

*A wise man learns from
experience and an even
wiser man from the
experience of others*

PLATO 428–348 BC

Contents

1	**CNC Controllers and Programming Techniques**	**1**
1.1	Introduction	1
1.2	CNC Controllers – a Review	3
1.3	The Sequence Used to Generate Part Programs	5
1.4	The Fundamentals of CNC Programming	8
1.5	High-speed Milling Fundamentals	76
1.6	High-speed Turning Operations	81
1.7	"Reverse Engineering" – an Overview of Digitising on Machining Centres	82
1.8	Computer-aided Design and Manufacture	89
Appendix	**National and International Machine Tool Standards**	**101**
Glossary of Terms		**102**
Selected Bibliography		**125**
Company Addresses		**127**
Index		**135**

Preface

Each volume in this series of three attempts to explain the design of turning and machining centres and how they are operated through part programming languages. Furthermore, a discussion about how such stand-alone machine tools can be networked into flexible manufacturing systems is given along with the problems relating to such interfacing. These volumes were written as a companion book to the successful *Advanced Machining – The Handbook of Cutting Technology* published jointly by IFS and Springer Verlag in 1989. The individual volumes look at interrelated aspects of using turning and machining centres:

Volume I considers the design, construction and building of turning and machining centres, then goes on to consider how these individual machine tools can be networked together providing the desired communication protocols for flexible manufacturing systems, leading to a complete Computerised Integrated Manufacturing system. This latter philosophy is discussed in terms of a case study on the most automated factory in Europe, ironically manufacturing turning and machining centres. Finally, mention is made of the efforts given to ensure significant advances in both ultra high-speed machining design and sub-micron operation, which is sure to have a major impact on general turning and machining centres in the future.

Volume II discusses the crucial point of ancillary activities associated with these machine tools, such as the cutting tool technology decisions that must be made in order to ensure that each machine is fully tooled-up and optimised efficiently. A brief review is also given on cutting tool materials development and tooling geometry considerations. Modular quick-change tooling is reviewed together with both tool and workpiece monitoring systems. A discussion follows on tool management, which becomes a major activity when a considerable tooling inventory exists within a manufacturing facility. Cutting fluids are an important complement to cutting tools, as they not only extend tool life but additionally enhance the workpiece machined surfaces; therefore it is important to choose the correct cutting fluid and handle it in the approriate manner to obtain maximum benefit from its usage. Workholding technology is an expensive burden that requires careful consideration to achieve an economic optimisation, particularly in a larger scale automated facility such as in an F.M.S. environment and a range of workholding strategies and techniques are reviewed.

Volume III is a highly focused text that discusses how a part program is generated – after a general discussion about controllers. Consideration is given to the fundamentals of CNC programming and this becomes a major part of the volume with a structured development of how to build programs and where and when the "word address", "blueprint/conversational" and "parametric" programs are utilised. High speed machining fundamentals are considered along with the problems of servo-lag and gain for both milling and turning operations. A section is devoted to "Reverse Engineering" using digitising/scanning techniques – allowing replicas to be used to generate part programs, as these techniques are becoming popular of late. Finally a discussion ensues on the design of CAD/CAM systems and how they might be used for multiple-axis machining, through a direct numerical control link.

Graham T. Smith
West End
Southampton
January 1993

Chapter 1
CNC Controllers and Programming Techniques

1.1 Introduction

The development of machine tool hardware since the introduction of numerical control in the late 1940s and early 1950s, has been progressively evolving and continues to do so, as we saw in chapter 1, Volume I. Even more dramatic than this development are the rapid advances in software engineering and electronic integration during this time and the speed of change has not abated. In fact with the advent of CNC in the 1970s, controller sophistication has considerably increased. The advantages gained from utilising CNC technology over conventional manual skills were argued in chapter 1, Volume I. The maturity of the latest controllers, in terms of their programming ability and reliability, makes them even more necessary for fast "turn-around" of parts coupled with improved productivity and flexibility in accommodating design modifications on components with the minimum of disruption to production.

The research and development departments where such systems are designed are continually looking for ways of simplifying controls and machine interfaces. System specialists are attempting to widen the scope of CNC and move into areas where its application would have previously been impracticable. The current machine controllers offer greater ease in their use – often termed "user-friendliness", a much abused term – with some versions allowing "conversational language program" (CAP) input in plain English, as an alternative to the conventional "word address format", yet to be discussed. As we shall see, with the more sophisticated and expensive CNCs, fully integral programmable logic controllers (PLCs) having a 32-bit data processing capability are increasingly being offered. The advent of Manufacturing Automation Protocol (MAP) and other Local/Wide Area Networks (LAN/WAN) has allowed previous stand-alone machine tools – often termed "islands of automation" – to be successfully interfaced to other machines, or peripheral devices typically associated with Flexible Manufacturing Cells or Systems. Such communication ability allows the successful data transfer of information from one controller to another, without corruption – often euphemistically termed "hand-shaking" – but more will be said about data transfer and communication techniques in the final chapter. Most systems are designed for "modularity" and "compatibility" – this latter term we have briefly touched on above. "Modularity" of the hard/software, means that the system/

machine tool designers have greater design flexibility, whilst an expansion in size and power of the integral PLC can accommodate both soft/hardware additions within the confines of the controller.

Most of the current controllers allow the programmer a variety of means of programming parts, such as:

"word address programming" – where a series of alpha-numeric characters, "G" and "M" codes associated with numbers, are used to control and manage the numerical distances between one feature and another

"parametric programming" – uses "free-variable" values which can be assigned to linear distances for either motions or calculations of slideway moves to obtain a component feature within the program. Such "parametric" techniques allow skeletal programs to be developed for a range of parts and by changing assigned numerical values, a feature can be changed at will, it can be used for a "Group Technology" (GT) approach for families of parts. "Parametric programs" can be condensed into fewer "blocks" of information for a desired part, when compared with "word address" formats and as such require less memory space (see section 1.4.10)

"conversational programming" – or as it is sometimes termed "blueprint"/"shop floor" programming – is achieved by answering a series of questions related to the desired part feature, e.g. pitch circle of holes, which is then compiled into a format acceptable to the CNC for tool motions. NB: A "graphics capability" is required to build up the desired features of the part and for "tape prove-out" (see section 1.4.11)

"background"/"parallel programming" – allows an operator to program a different part from one that is at present being machined – using a different area of the memory of the CNC for its compilation. This considerably speeds up the lead-time necessary for the production of a successive part

"off-line programming" – is usually undertaken on either a CAD/CAM workstation, or using "computer-assisted part programming" (CAPP) equipment normally with a "direct numerical control" (DNC) link to the controller through the RS232 interface, or via punched paper tape and tape reader into the controller – rather an old-fashioned method nowadays. NB: "Off-line programming" can be achieved away from the machine tool. This overcomes the distractions associated with programming in a production environment and as such, allows the programmer to develop programs under ideal conditions whilst not affecting the demands of production

"digitising" – often termed "reverse engineering", whereby a model or component's dimensional features are "captured" and used as a basis for the generation of a part program. NB See Figs. 1.32 and 1.33, together with a description of the process, in section 1.7.

In the following sections we will see how the use of programming aids can minimise the length of programs. This may be an important criterion when one is faced with either a small memory capacity in the CNC, or many parts already stored in the memory, or a complex part geometry, which might otherwise take up considerable space in this memory. These programming aids are usually available for most CNC controllers, or may be offered as "options" from many universal/proprietary builders. Such options are: "mirror-imaging", "scaling", "angular rotation", "datum shifting", "canned-cycles", "subroutines" and "nested subroutines", which all reduce the length of the part program considerably, with the secondary benefit of minimising the expertise necessary in writing such programs. The structure of "word address programs" will be enlarged upon later – arguably the most common technique used in developing CNC part programs for both turning and machining centres. For the sake of clarity this topic will be based, in the main, upon one controller made by a popular universal CNC builder. It should be said that there is not enough space within the text to devote to an exhaustive account of how such part programs are developed, but an appreciation of the main aspects of the logical structure and compilation of a program is all that is anticipated from the reader. To gain more insight into and appreciation of the unique problems associated with programming/operating machine tools, specific courses and their respective manuals are available when either CNC machines are purchased, or new personnel require retraining. Such courses give the would-be programmer a measure of expertise in part program development/manipulation and its subsequent operation for given part geometries. Any specific CNC machine tool offers a unique intellectual challenge to the programmer which must be overcome in order that satisfactory part programs are developed.

Prior to a discussion on the part programming philosophy and strategy needed to produce successful programs, there will be a brief review of just some of the typical controllers currently available from both universal and proprietary builders. It is hoped that this will give the reader a greater insight into the subject of CNC controllers and aid in the understanding of later sections in the chapter.

1.2 CNC Controllers – a Review

In order to discuss the diverse range of controllers available for CNC machine tools, it is necessary to try and highlight the differences with just a few of those currently available. One of the most popular controllers is made by GE Fanuc. It offers a highly sophisticated design and programming ability. This high-speed controller was first shown at the "7 EMO" exhibition in Milan in October 1987. It became available to customers during 1988 and by the 1990s has seen universal acceptance across a large range of machine tool builders. Its principal features are that owing to its high processing speed, it can be used for complex die sinking, or high-speed machining operations. It also offers a number of artificial intelligence (AI) features, allowing it to make logical decisions in concert with changing process needs, including: simple "conversational programming" and guidance in intelligent failure diagnostics.

By using multiple 32-bit high-speed Motorola 68020 microprocessors, together with the original 32-bit multi-master bus, it can achieve machine command processing rates up to ten times faster than most currently available CNCs. Furthermore, an ultra high-speed programmable machine controller allows interface commands to be executed up to eight times faster than previous systems. With such advanced microelectronics comes a controller which has been significantly reduced in size, principally because of using surface mount and customised large scale integration (LSI) electronic component technology. Reliability is another bonus of such technology integration, whilst a high-speed backplane bus architecture allows its boards to be utilised in virtually any number and combination. Such a system has a fully digital interface to high performance AC servo-control technologies, giving a quick-response servo-system that is unaffected by mechanical load variances – this means that with fully electronic absolute position detection, machine referencing is made a thing of the past.

For high-speed machining capability, this controller can command multiple axes in both rapid traverse and cutting speed up to 100 m/min with a resolution of 1 µm, or alternatively, traverses at 24 m/min with 100 nm resolution. Speed of machining can be increased still further if the optional 32-bit subprocessor is used, permitting super high-speed machining of continuous short blocks. For example, if the requirement is for 3-axis simultaneous DNC operation in 1 mm continuous blocks, machining speeds of 15 m/min in EIA format, or 30 m/min in binary format input can be achieved. This requires the RS 422 interface to process data up to a maximum speed of 86.4 kbps – reading the high-speed information from the host computer. Such simultaneous control necessitates processing capabilities of the highest order, in order to minimise "shocks" to the machine as it: starts, stops, accelerates, and decelerates. Therefore a "look ahead" capability of up to 15 blocks of command information is necessary and a smoothing of acceleration/deceleration occurs.

Much more could be said about such a controller, mentioning such features as its ability to make automatic decisions of tools and cutting conditions, the arbitrary tool path editing function, tool interface checks with on-screen animation, automatic process determination – creating machining processes to cut parts if the blank and part figure have been defined, together with a variety of interpolation modes: linear, circular, cylindrical, involute, spline, polar coordinate, exponential, spiral, circular thread, and finally hypothetical axis interpolation. However, this would entail a chapter alone, although several such advanced features will be discussed later in this chapter.

The Siemens Sinumerik controller offers a competitive alternative to the one described above and both are available, when suitably configured, for either turning or machining centres with multiple axis capability. The Sinumerik universal controller utilises a range of "soft keys" situated below the screen, enabling the operator to choose which menu to monitor. In the previous controller more "soft keys" allow considerably more functions to be chosen as the operator desires. Any of the currently available sophisticated controllers have the capability of compensating for minute variations that occur in the manufacture of the leadscrew. Such leadscrew error compensation must be achieved in all axes simultaneously and in the Sinumerik up to 1024 compensated positions in all axes are possible. The distance between all compensation positions is selectable in the range 0.01–320 mm on any axis, whereas the resolution can vary 1–64 µm. Such errors are determined by laser interferometry for each linear axis as explained in chapter 1, Volume I, but may be modified at periodic intervals when the machine tool is recalibrated.

Both of the previous controllers can be termed "universal" and as such may be purchased for a variety of machine tools and configured accordingly. However, some controllers are manufactured by "proprietary" machine tool builders and are generally not sold for incorporation onto other company's machines. The Electropilot (Gilde-meister (UK) Ltd) controller has been built with a totally different philosophy. What is particularly noticeable about this controller is the absence of many of the "keys" associated with the other controllers, making it popular and very easy to operate on the shop floor, but still having a high level of computing sophistication. This modular multi-processor system utilises 16-bit microprocessors, supplemented as necessary with 32-bit units, coupled to specific processors for axis control. This allows the controller to be expanded from 2- to 4-axis machines catering for: rotary axes, interfaces to DNC, linking into larger manufacturing systems, or with the host computer.

The Acramatic controller (Cincinnati Milacron) is a "dedicated" universal controller which can be fitted to either machining or turning centres. Yet again it has many of the features described on the more sophisticated "universal" controllers mentioned above, but it has been "customised" for a specific application. A novel feature of its 4-axis cousin, is that when programming part features, one need only worry about the relative motions in two axes, namely "X" and "Z". Once the program has been compiled and proven, it automatically selects operations using the second turret for either simultaneous machining such as: "balanced" turning, drilling and turning, boring and screwcutting, and so on, simplifying multi-axis turning operations still further.

A major problem either overlooked or rarely addressed by companies looking for a machine tool, is the time and number of "key strokes" necessary to input a program successfully into the controller's memory. This is particularly relevant when a company must write and "prove-out" the programs at the machine tool. If many "key strokes" are necessary to write a program, each one is an opportunity to produce an error, requiring further editing under the distractions of a shop floor environment. Comparison between the amount of "key stroke" processing and the time it takes to input the same program is a good "bench mark" of its "user-friendliness" and programming logic. Any company considering the purchase of a CNC machine tool, will not find this a wasted effort during any feasibility study undertaken on their "short-list" of machines.

To complete this review of CNC controllers, the latest top-of-the-range Acramatic controller (Cincinnati Milacron) is worthy of mention. It has an almost word processor-like capability allowing "cut and paste" editing of part programs, so that single characters, blocks, or strings of characters can be defined, erased, copied, or inserted quickly and easily. Probably its most unique feature is the touch screen with pop-up windows and a calculator which provide easy function selection with less risk of error. The controller offers a real-time multi-tasking operating system in which nine microprocessors share in performing a range of tasks, optimising machine productivity and part quality, whilst the operator is involved on background tasks – such as downloading a program, or tool data revision. Whenever demanding applications are the requirement, such as running at maximum processing speed and accuracy, this necessitates the controller having optional 32-bit Intel 80386 processors.

A multiple set-up support package makes this controller ideal for any untended operations, providing scheduling, sequencing and status definition capabilities usually available only in workstation managers and similar off-line systems. As well as pallet and part scheduling together with sequence definition, users can define the pallet, fixture, and part offsets up to a maximum of 16 pallets of parts, but with sufficient offsets available to accommodate up to 64 parts for these pallets. With such a comprehensive "pallet pool", the tooling database, of necessity, must be large and the tool tables have a separate program memory, allowing users to manage up to 500 tools, with the expanded optional memory. Additional features include over 20 000 metres of part storage on a 20 megabyte hard disk, with 32-character alphanumeric program identifiers, coupled to RS 491 level II data line and terminal emulation capabilities. Such controllers are becoming "industry standards" for machine tools, offering greater flexibility, sophistication and adaptability to both operator and programmer alike.

1.3 The Sequence Used to Generate Part Programs

Prior to writing a part program for a machine tool there are many important considerations which must be addressed, if the part is to be manufactured successfully. It is simply not just a matter of stating the component geometry, or even considering the motions of the tool's path, without taking into account such crucial factors as how the part is to be held and on what machine, which tools should be selected and cutting data utilised. Considerable skill is necessary if these inter-related decisions are going to produce a component that is "right first time".

If we consider the sequence necessary in the generation of a CNC program, it can be thought of as four fundamental stages:

problem description
processing
control
adaptation

From Fig. 1.1 one can gain an appreciation of the many inter-related activities which go towards the successful completion of a part program. In this example, the part is either defined by a CAD/CAM system, or developed using a programming language such as "word address" at the formative "problem description" stage. The geometric data for the blank and finished part may include sub-programs together with the cutting sequence within its main program, giving the programmer a process description that can now be "processed". At this stage, assuming that the initial part program is acceptable, the details such as specific tooling requirements, workholding and materials can be defined. It is also assumed that a machine tool has already been chosen which will be compatible with the expected production demands for the part, in terms of both quantity and quality required. Knowing the machine tool specification can allow a degree of flexibility in our calculation of the cutting sequence and the distribution of cuts, cycles and axes chosen. For example, during this "processing" stage and assuming that a 4-axis turning centre is specified to manufacture the parts, we might want to "balance turn" the part, or machine features using the C-axis – perhaps with "driven" tooling. Decisions are made to produce the part in the most efficient manner and at this stage "collision monitoring" problems can be explored through the use of a graphical display of the cutting sequence. This will ensure that collision of both tools and workholding equipment is avoided in the dynamic display on the screen; furthermore this output can be plotted for a "hard copy" file, for later use/verification.

Finally, once the programmer is happy that all these conditions have been met and any anticipated problems resolved, the final CNC program is produced together with the necessary time calculations and a copy is stored in either the program library, perhaps using a DNC link, or for permanent hard copy storage on an NC punched paper tape, together with the complementary setting sheets, times, etc. It is important to ensure always that a "back-up" copy of any part programs is safely stored away from the machine tool in some secure and fire-proof environment. This is because it is quite easy to over-write and erase a program left in the machine tool memory inadvertently, or lose the programming sheets kept within a planning office, with no particular tape library facilities. It should be remembered that a company's time, cost and expertise are tied up in such programs and they should be treated as a valuable resource, worthy of protection.

In the following sections we will explore how the structure of a CNC program is produced and go on to consider how this relates not only to the machining and turning centre's datums, but mention the methods of choosing coordinates relating to the workpiece and some of the special features allowing the programmer greater ease in defining the part, whilst minimising its length. Later on, other sections will look at specific cases of interpolation methods and actual part programming problems, as well as the effects of cutter compensation and the use of "canned cycles". Finally, applications such as high-speed machining methods and digitising techniques will be referred to, followed by CAD/CAM solutions, in producing complex profiles on components.

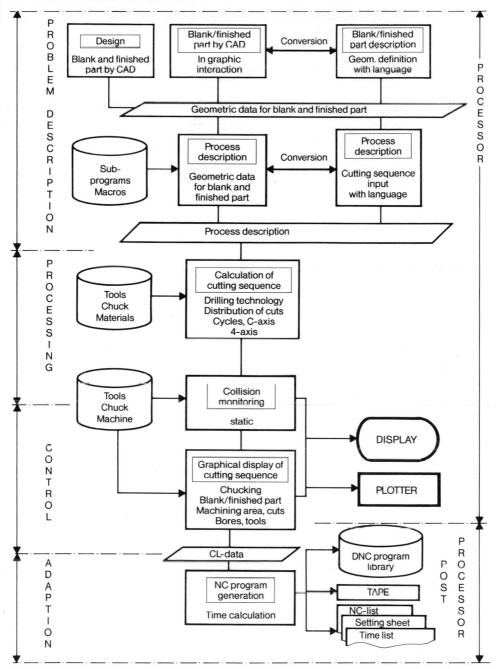

Fig. 1.1. Generating a sequence for a CNC program. [Courtesy of Gildemeister (UK) Ltd.]

1.4 The Fundamentals of CNC Programming

By now the reader should appreciate that any CNC machine tool is guided by its part program through the controller. In order to perfrom the necessary machining operations, the controller needs specific information such as:

workpiece dimensions
tool travel and the axis of the slideways
machining sequence
tool selection
speeds and feedrates

The program will sort this information into the necessary sequence and then translate it into a language which can be understood by the controller. In the following programming examples they have been written in an accepted ISO code, with all geometric/linear values in metric units. The programming instructions describe just the standard range of functions of the system, with the maximum values specified being limit values and during operation they may be restricted by machine data, interface unit and input/output devices.

Program Structure

The programming examples throughout this chapter are based on the DIN 66025 structure.

Any part program is comprised of a complete string of blocks which define the sequence of operations for a machining process on a CNC machine tool. The part program (Fig. 1.2a) comprises:

the character for the program start
a number of blocks
the character for the program end

The character for the program start always precedes the first block in the part program, whereas the program end character is obviously contained in the last block.

If we consider the part program structure in terms of its input/output format, one may use a mixture of "subroutines" and "canned cycles" as being components of the program. As was mentioned earlier, more will be said on these topics later in the chapter, but for now it is worth mentioning that such "canned cycles" and "subroutines" are normally generated either by the machine tool builder, or the controller manufacturer – assuming it is made for universal use. Most standard program memories can store about 200 part programs and "subroutines" simultaneously, with their input sequence being purely arbitrary. Usually, whenever the programmer enters a part program manually, termed "manual data input" (MDI), with the "block number soft key" activated, any future "block numbers" are generated automatically with either steps of 5 or 10 between them, or as on some controllers, step numerical values can be varied, from the first block. The "cancel" key can be used to delete any entered block number, or the "edit" key could be used to overwrite this block's information.

Block Format

A block contains all the data necessary to implement an operating procedure, which can contain several words and the "block end" character (see Fig. 1.2b). There is

a PROGRAM STRUCTURE.

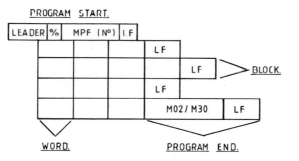

b BLOCK FORMAT.

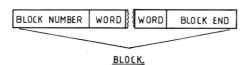

c WORD FORMAT.

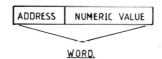

d EXTENDED ADDRESS.

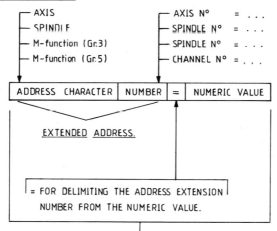

Fig. 1.2. The fundamentals of programming, based on DIN 66025. [Courtesy of Siemens.]

usually a finite limit to the length of block and this might typically be around 120 characters; furthermore, the block is displayed in its entirety over several lines, when approaching the maximum value. Any block is entered under an address "N" with block numbers being freely selectable. Defined "block search" and "jump" functions can only be guaranteed if a block number is used no more than once in any program. It is also permissible to program a block without a block number and in this case it is not possible to either "block search" or use the "jump" functions. Obviously any block format should be made as simple as possible, by arranging the words of the block in the sequence of the program key, as shown in the block example below. NB The controller has a "default condition" where "G" and "M" codes are automatically preset upon controller at start-up and must be over-written when changes are required, such functions are "modal".

N6832 G..X..Y..Z..F..S..T..M..LF
where
N = address of block number
6832 = block number
X..Y..Z.. = position data
F = feedrate
S = spindle speed
T = tool number
M = miscellaneous function (see section 1.4.8)
LF = block end

Each block must be terminated by an end-of-block character "LF". Such a character may/may not appear on the screen, or its equivalent, depending upon the controller, although when a program is printed out it is usually omitted.

Block Elements

There are two types of blocks designated in most controllers:

main blocks
sub-blocks

With the "main block", it must contain all the words that are necessary to initiate the machining cycle in this section of the program. A main block is normally identified by means of a ":" (colon) character, instead of the address character "N", as depicted below.

:20 G01 X 15 Y − 20 F 250 S1100 M03 LF

Conversely, a "sub-block" contains only the functions which are different from those in the previous block:

N12 Y40 LF

A main block and several sub-blocks can constitute a section of a part program:

:
N210
N220 } Section
N230

At this stage is our discussion of the various block formats, it is important to realise that the "preparatory" and "miscellaneous" functions, together with other "words" may be designated as either

modal
non-modal

A "modal" function remains active until it is either cancelled, or superseded by another complementary "word", e.g. G-code for a G01, whereas a "non-modal" word is only active in its particular block. Such a program structure reduces the repetitiveness in programming whilst simultaneously simplifying and reducing program lengths.

Deletable Blocks

Whenever program blocks must not be executed during every program run, they can be "skipped" by entering some character such as a "/" in front of the block number with the word. Such deletion of blocks is activated via either the machine control panel, or the interface controller. Any deleted block/s must form a loop, with the start and end at the same point, or the program may be executed incorrectly, although a section can be "skipped" by deleting several consecutive blocks.

Word Format

A "word" is an element of a block, comprising an "address character" and a string of digits (see Fig. 1.2c). The address character is usually a letter, with the string of digits being specified with a sign and decimal points. Such a sign "+/−" value is written between the address letter and the string of digits, although the positive sign can be omitted.

Extended Address

The structure of an "extended address" is depicted in Fig. 1.2d, simply as an example of the sophisticated designations and subtle distinctions that can be made within a block. However, owing to the complex nature and variations of such an address, it is not possible to describe, in the space available, its use in the detail necessary, so further discussion will not be given and the reader should refer to specific programming manuals for an actual CNC controller, if more information is required.

1.4.1 Machine Tool Reference Points

Any CNC machine tool has a means by which each axis can be referenced with respect to a known datum. Normally several axis delimiters exist (i.e. datums):

software stops
hardware stops
   (a) micro-switches, or similar
   (b) mechanical buffers – preventing axis over-travel

When a machine does not have a "fully digital interface" with advanced software control, it will need to reference the axes prior to use, then a "home" position-datum must be known to ensure that any motional travel under CNC control is related to such a reference. Confirmation that "home" has been established for each axis is necessary and usually either lights or LEDs for each controller axis will confirm that the software stop position has been reached, after a short "hunting" motion.

Of course, the arrangement of such origin positions, datum points, will differ for different machine tool configurations, which is related to the machine's coordinate system. Two such systems are illustrated in Fig. 1.3, for turning and machining centres, and it is apparent that in both cases several datums are present. Obviously, a machine zero reference position, or "home" occurs, for the establishment by the machine tool manufacturer of the origin for all axes, which synchronises the system. Other datums which must be established are for the workpiece reference point and zero offset, together with the tool reference point. The workpiece zero is defined for programming the workpiece dimensions and it may be freely selected by the programmer. The relationship of this datum to the machine zero is defined as its zero offset in one or more planes. The datum point for the tool setting positions cannot be varied in one aspect, namely the gauge line for a tool on a machining centre, or the turret's centreline for a turning centre, but their reference point coordinates can, of course, be modified.

Many CNC controller builders offer a range of other offsets, selected by suitable G-functions, such as settable zero offsets, or programmable zero offsets for the workpiece.

1.4.2 Types of Coordinates – Dimensioning Systems

The traversing movement to a particular point in the coordinate system can be described by means of three types of coordinates:

absolute dimensional positions – data input "G90"

incremental dimensional positions – data input "G91"

polar dimensional positions – data inputs either G00, G01, G02, G03

If "absolute" position data input is selected (Fig. 1.4a), all the dimensional inputs refer to a fixed zero, which is usually the workpiece zero. The value given to associated position data specifies the target position in the coordinate system.

When "incremental" position data input is selected (Fig. 1.4b), the value of the position data corresponds with the path to be traversed. The direction of axis move-ment is specified by the positive/negative sign.

NB: It is possible to switch between absolute and incremental position data input, from one block to the next as desired, since the controller's actual value is always referred to the zero. Furthermore, a zero offset is calculated for both the absolute and incremental programming of part features.

With many control systems, it is possible to program part features using "polar" position data input (Fig. 1.4c); however, there is a variety of methods of machining polar features. Using point-to-point, linear, or circular interpolation motions, polar features can be developed, but to differentiate polar from rectangular coordinates, particular word address commands are used. If, for example, a series of holes on a pitch circle diameter is to be drilled, the centre of these holes (known as the "pole") must be established in the plane to be machined. If we assume that the X and Y plane

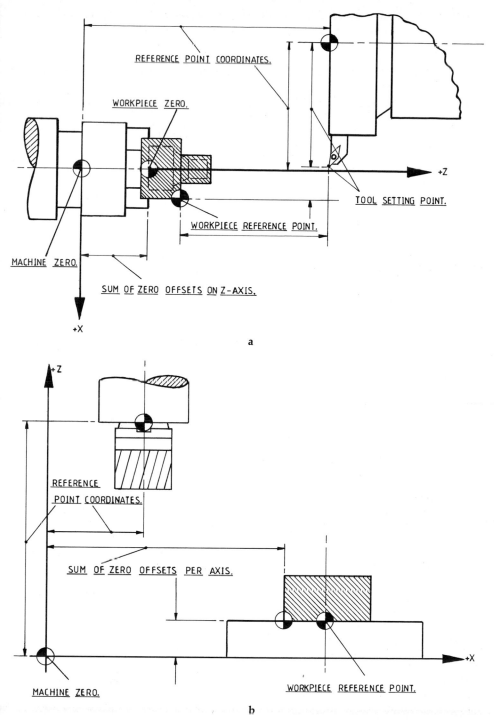

Fig. 1.3. a The reference points for turning centres (basic configuration). **b** The reference points for machining centres (basic configuration).

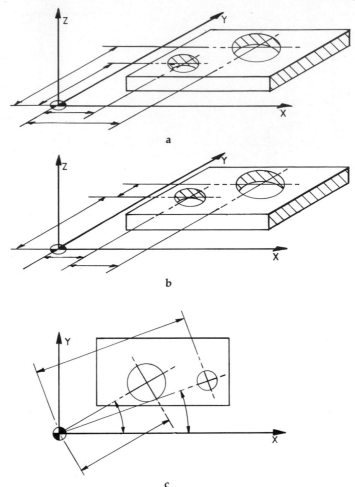

Fig. 1.4. a Absolute coordinates. **b** Incremental coordinates. **c** Polar coordinates.

is utilised with positive signs, then the linear distance from our workpiece zero in the "X-axis" is denoted by an "I" character and in the "Y-axis" by a "J" character. Two more word address characters must be designated to confirm the first hole position to be drilled, the radius of the hole given the character "R", which is always a positive sign, and the angular position of this hole, denoted by the character "A". Further holes to be drilled – possibly using "canned cycles", of which more will be said later in the chapter – simply require their angular positions to be confirmed in the relative blocks; although if the holes around the pole differ in radial values, then both the angular and radial characteristics, "A" and "R", must be stated for subsequent holes to be drilled. The characteristics of "I", "J", "R", and "A" will differ only marginally if, for example, the "X" and "Z" planes were chosen and a "K" character substituted for the "J", with the others being the same designations. Later on, a more detailed description, together with programming examples of a linear, circular, helical, and cylindrical interpolation is explained in section 1.4.4.

1.4.3 The Cutter Transformations of Angular Rotation, Mirror-imaging, Datum Shifting and Scaling

Such programming aids, when available in the controller's logic, offer several benefits to a programmer in the work required to execute a feature, whilst compressing the program into fewer blocks of information and as such, minimising memory space. These programming aids, termed "cutter transformations" are used because they have the ability to transform a cutter path from the current programmed position to another.

Let us begin by examining the "angular rotation" cutter transformation and then go on to consider the benefits to be gained from the others in turn, namely "mirror imaging", "datum shifting", and "scaling".

Angular Rotation

Whenever a geometric feature such as a milled contour, or drilling pattern needs to be rotated around a fixed position, then the "angular rotation" cutter transformation can be used as illustrated in Fig. 1.5. When such features as impeller blades need to be machined in a fixed angular relationship to each other then we only need to be concerned with cutting one blade entirely when programming the cutter motions and relying on the angular rotation function to execute the remainder. Polar translation is closely allied to rotation and permits a programmer to reposition a feature with respect to a centre/pole, allowing the feature to be rotated about this predetermined centre. An added bonus of such translation is that complex-shaped features requiring angular positional location may be programmed as if they were orientated to the true "X-Y" axes – which simplifies calculations.

Mirror-imaging

The "mirror-imaging" cutter transformation (Fig. 1.6) permits the machining of a contour, or hole pattern, by the inversion of one or more axes, enabling the feature to be machined in another plane. Such "mirroring" of a coordinate axis permits contour machining/drilling in the following relationships:

with the same dimensions

at the same distance from the other axes

on the other side of the "mirror axis", but as a "mirror-image"

During the "mirroring" sequence the controller inverts:

the sign of the coordinates of the mirrored axis

the direction of rotation, in the case of circular interpolation, i.e. G03 to G02, or vice versa

a machining direction, i.e. G41 to G42, or vice versa

Whilst "mirroring" is active in the part program, there is no "mirroring effect" on either the tool length offsets, or the zero offsets on a machining centre. Conversely, in the case of a turning centre, the following condition will be "mirrored" on the "X-axis" – the position of the tool cutting point – although when "mirroring" on the "Z-axis" this does not apply. A "mirroring" operation is always in relation to the

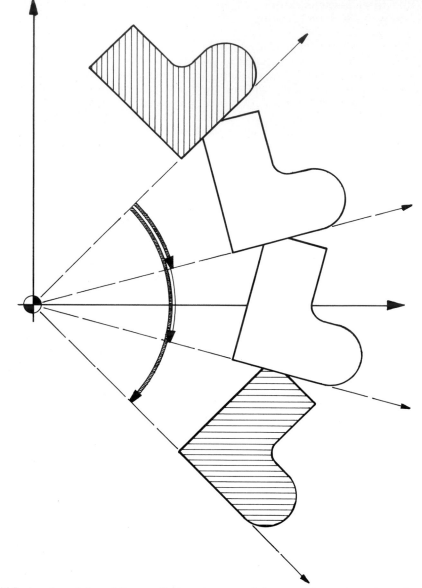

Fig. 1.5. Angular rotation of the coordinate system around the actual datum, i.e. either "absolute" or "incremental".

coordinate axis so that contours, etc., may be "mirrored" in the exact position where they are required to be machined. The position of the program start necessitates the "mirror-call" such that the axes of the coordinate system are located exactly between the programmed contour and the "mirrored" contour. In order to achieve this axis symmetry, the zero of the coordinate system can be offset to the correct

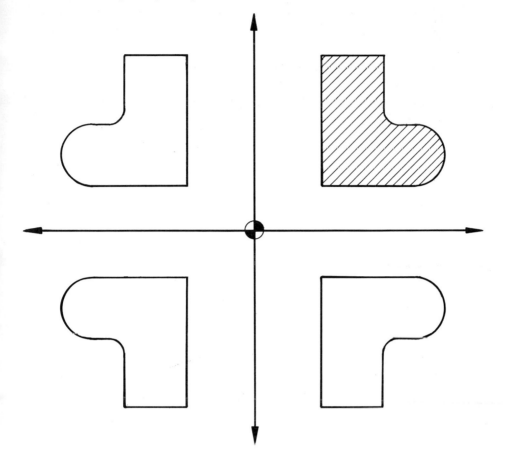

Fig. 1.6. Mirror-imaging of milling/drilling/boring patterns in one or both axes of the maching plane.

position, before any "mirroring" call in the program is activated – normally using "M-functions".

NB: A controller manufacturer has set distinct rules for the engagement of such cutter transformations and readers need to familiarise themselves with the specific programming manual to gain a greater insight into their execution.

Datum Shifting

The cutter transformation known as "datum shifting" or "preset absolute registers" is, in the main, used in milling operations, but it can also have applications for turning. However, in this case, we will only consider its use for milling/drilling applications. Probably the most often used preparatory function for "datum shifting" is "G92" – shown schematically in Fig. 1.7. It enables the programmer to "shift" the zero datum to any position within the main machine tool's envelope in the X, Y and Z planes. These "shifts" are of a temporary nature, lasting only whilst the part is

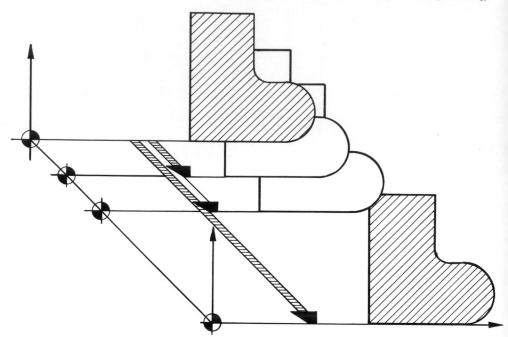

Fig. 1.7. Datum shifting of the coordinate system. i.e. either "absolute" or "incremental".

being machined and either are reset to the original datum, or "shifted" to a new one. The purpose of such "shifts" of datum positions can be threefold:

to cut a range of identical parts out of wrought stock – a technique favoured by the aerospace industry for aluminium and its alloys, and described in chapter 4

for machining identical parts held on a multiple fixture and, in this case, the "datum shifts" allow the same part to be machined with little additional increase in block lengths – usually as a "sub-routine", but more about this technique later in the chapter

machining different parts held on multiple fixtures – but differing from the example given above after the "shifts" have been "called": "sub-routines" for each individual part to be machined are "called" until all the parts have been cut and the datum is reset to the initial position

Scaling

"Scaling" is a very helpful cutter transformation when used, in particular, for milling contours and to a lesser extent for drilling operations. A typical schematic of the "scaling" of a contour is given in Fig. 1.8, illustrating the range and size variation that can be achieved by using "scaling factors". Scaling allows not only internal and external features that are identical to be machined – "capturing", for example, the outside profile and using it as a "sub-routine" which has been appropriately "scaled" to give an identical but reduced internal profile – but it can also be used to change,

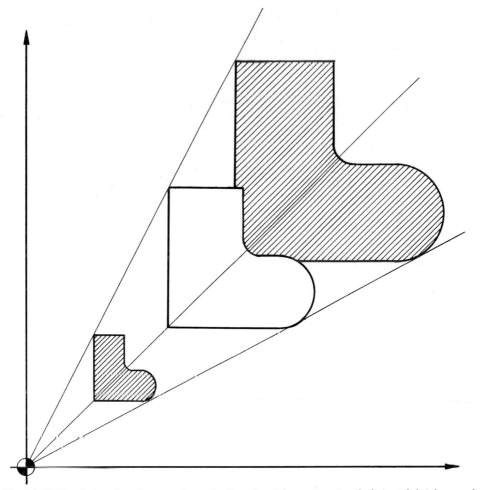

Fig. 1.8. Scaling factor: the enlargement or reduction of contours, automatic calculation of shrinkage and growth allowances, etc.

say, a circular interpolated shape, or spiral into an elliptical profile by "scaling" one axis. Such "axis compressions" of circular interpolated features give a considerable scope – outside the expected specification of the controller – and may be used to simulate parabolic interpolation. This means that once the tool path of a symmetrical profile has been generated, it may, at will, be compressed to achieve an almost free-form shape, which would otherwise only be expected from "captured geometry" sent to the controller via DNC or equivalent, through a CAD/CAM workstation. It is very easy to "compress" one axis using the correct preparatory "G-code" at one instant, then cancel the "scaling" and compress another axis either within a "skeletal" part program, or in future blocks as necessary.

Not only is it possible to use this technique for producing a range of products in a "group technology" approach, similar to "parametric programming" – yet to be described – but if a contour is large or the program is complex, by reducing its size and using a high "feedrate-override", tape prove-out times can be considerably reduced. The "G-codes" used for "scaling" and their cancellation are at the discretion of the controller manufacturer and the rules for their implementation/engagement are stated in the relevant programming manuals.

NB: When an axis is "scaled" – or indeed both axes for part symmetry – this also means that the component's relative position "drifts" (see Fig. 1.8). This requires the use of a "datum shift" in combination and proportion to the scaled axis, or axes, in order that it is machined in the correct position/orientation to the detailed drawing requirement.

1.4.4 The Programming of Motion Blocks

In chapter 1, Volume I we saw that each axis – whether primary, secondary or tertiary control – is designated as either a linear or rotary motion. Each axis motion can be controlled and programmed at will as a "rapid" or a "feed" motion and in this section we will discuss such motion control in more detail along with interpolation techniques, together with the preparatory functions needed to successfully engage them.

Axis Motion without Machining – Using Preparatory Function G00

If rapid motions are required then they are programmed by means of the position data G00 – with some controllers one or both of the zeros can be omitted – together with a target position specification. The target position is reached using either an absolute position data input (G90), or an incremental position data input (G91). Any path programmed using the G00 function is traversed at the maximum possible speed (rapid traverse) along a straight line by linear interpolation, without machining the workpiece. The maximum permissible speed of an axis is monitored by the controller and is defined as machine data for the axis. If simultaneous rapid traverse is required in several axes, the traversing speed is determined by the lowest axis speed specified in the machine data.

When the G00 preparatory function is programmed, the feedrate previously "active" under an "F-word" will be stored but can be called again by initiating a G01 function.

Axis Motions with Machining

A controller will implement either linear or circular interpolation depending on the type of axis motion required, as shown below:

linear interpolation – produces linear motion by paraxial moves in either two or three axes

circular interpolation – produces a circular motion in a plane by the synchronised movements of two axes

Linear Interpolation

This preparatory function (G01) allows the tool to travel at a set feedrate along a straight line to the target position, whilst simultaneously machining the workpiece.

The controller will calculate the tool path by means of linear interpolation. Linear interpolation effects a motion:

in one axis direction, either a linear or rotary axis

from the starting position to the target position which can be programmed using either absolute or incremental position data

at the programmed feedrate

at the programmed spindle speed

An example of 3-axes linear interpolation can be seen in Fig. 1.9a, with word address program:

```
%10LF
N1 G00 G90 X40 Y60 Z2 S500 M3 LF
N2 G01 Z-12 F100 LF
N3 X20 Y10 Z-10 LF
N4 G00 Z100 LF
N5 X-20 Y-10 LF (or: X0, Y0)
N6 M30 LF
        N1 = tool rapid traverse to P01
        N2 = infeed to Z-12, feedrate 100
        N3 = tool traverse along a straight line in space to P02
        N4/N5 = rapid traverse clear
        N6 = end of program
```

Circular Interpolation

The tool traverses between two points on a contour in a circular arc, whilst simultaneously machining the workpiece. To achieve such cutter motion the controller must calculate the tool path using circular interpolation, based upon the following rules when machining along a circular arc:

in a clockwise direction with G02

in an anti-clockwise direction with G03

in the desired plane in the case of milling by a freely selectable plane with G16, or,

in the X–Y plane with G17 ⎫
in the Z–X plane with G18 ⎬ see Fig. 1.9b(ii)
in the Y–Z plane with G19 ⎭

around the programmed centre point of the circle

from the starting position on a circular path to the programmed end position

The preparatory functions of G02 and G03 are "modal" in action, with the circular motion being executed in any selected plane – XY, XZ and YZ. The rotation direction of the cutter in various planes is defined in the following manner – standing facing the axis which is perpendicular to the plane. The tool will move in a clockwise direction with G02, or in an anti-clockwise direction with G03. The interpolation parameters determine the circle or circular arc in conjunction with the axis commands, with the starting point at "KA" (see Fig. 1.9bi) being defined by the preceding block. The end position of the arc "KE" is defined in this case by the chosen axis values of "X and Z". The centre point of this arc is "KM" and may be defined in one of two ways:

by the interpolation parameters
directly using the radius

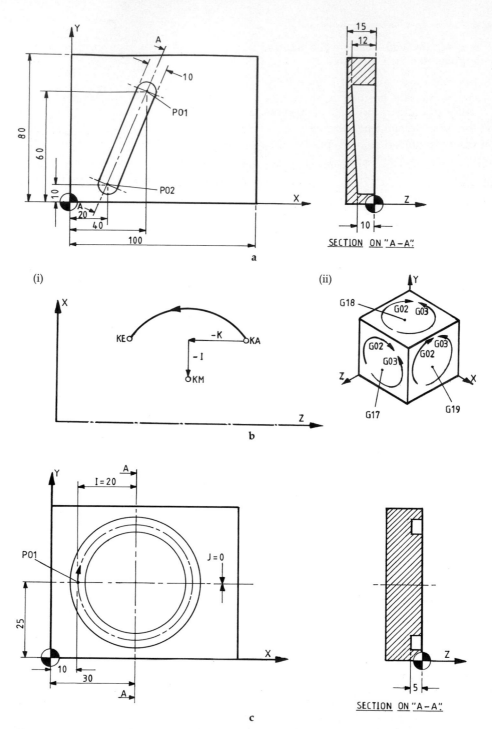

Fig. 1.9. Linear and circular interpolation applications, according to DIN 66025. **a** Linear interpolation in 3 axes. **b** (i) Circular interpolation with interpolation parameters. (ii) Circular interpolation. **c** Interpolating a full circle in the X–Y plane. [Courtesy of Siemens.]

Interpolation Parameters (I, J and K)

The interpolation parameters "I, J and K" are the paraxial coordinates of the distance vector from the starting position to the circle's centrepoint. In accordance with the DIN66025 standard for word address programming, the interpolation parameters "I, J and K" are allocated to the respective axes "X, Y and Z". Such interpolation parameters must always be entered as incremental position data, irrespective of whether the axes "X, Y and Z" are programmed using absolute, or incremental position data (see Fig. 1.9bi).

The sign chosen is based upon the direction of the coordinates from the starting point of the circle, or arc (Fig. 1.9bi). If a value of zero for the interpolation parameter is assigned, its sign is not programmed. Similarly, any end position coordinates which are the same as at the start point for the circular path need not be programmed – typically when generating a full circle – but at least one axis must be programmed, i.e. "X0, Y0 or Z0".

In the example shown in Fig. 1.9c, we can see that a full circle has been programmed into our rectangular block, with a typical execution of the part program shown below, but noting that according to the rules above, the start and end positions are the same.

```
%20 LF
N1 G00 X10 Y25 Z1 S1250 M3 LF
N2 G01 Z-5 F100 LF
N3 G02 X10 Y25 I20 J0 F125 LF
N4 G00 Z100 M5 LF
N5 X-20 LF
N6 M30 LF
```

N1 = tool rapid traverse to point P01
N2 = infeed to Z-5
N3 = X–Y plane automatically selected – i.e. a "default condition". Tool traverses clockwise around the full circle (G02)
N4/N5 = rapid traverse clear
N6 = end of program

If one incorrectly inputs the values for the interpolation parameters "I, J and K", the circle end position check will detect such programming errors – this is providing that they are not within the tolerance range. Under such conditions no circular interpolation results, furthermore, it is usual for some form of alarm to be displayed. Problems can arise, however, when the programming error is within the tolerance range and the end position of the circular arc will be approached exactly – by offsetting the centrepoint of the circle, but the tool path between the start and end positions will be as follows:

when the interpolation parameter is too high, undercutting of the circle occurs and the circle end position check may be suppressed

if the interpolation parameter is too low, extra stock is left on after the circle has been machined

NB: The setting range of a typical system for the tolerance and the circle end position is $\pm 1\,\mu m$ to $\pm 32\,000\,\mu m$ and this tolerance range can be compensated for by entering it as a value without a sign.

Radius Programming

For many applications, dimensioning drawings, using the radius, is a better method of specifying the feature rather than the diameter. By using such characters as "U" or "B" when programming it simplifies the task, as illustrated in Fig. 1.10a,b, for defining the circular path the tool must take when generating a contour. The programming logic dictates that since the radius to be machined is used in conjunction with either a G02 or G03 preparatory function, it can only specify a definite circular path within a semi-circle. Therefore it is necessary to specify additionally whether the traversing angle is to be greater or smaller than 180°. In order for the controller to discriminate between the various magnitudes of angle required, it is allocated one of the following signs:

+U/+B, for angles less than or equal to 180°
−U/−B, for angles greater than 180°

It is not permissible to program the radius by such methods if it has a traversing angle of either 0° or 360°. When full circular paths are demanded they must be

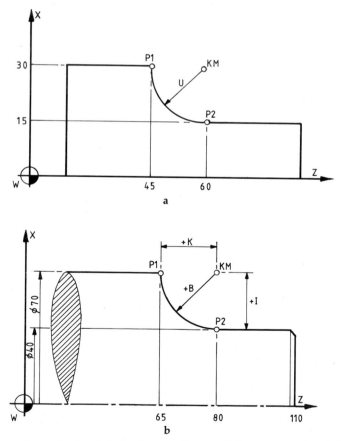

Fig. 1.10. Programming radius bends in turning operations. **a** Radius programming. **b** Circular interpolation. [Courtesy of Siemens.]

programmed utilising the standard interpolation parameters described earlier in this section. We have seen how G02 or G03 determines the direction of movement around the circle, which is defined by means of the circle end position and the interpolation parameters. In the examples below we can gain an appreciation of the same circular path generation, but in either case the radius may be defined using either a "U" or "B".

Example 1. Relating to Fig. 1.10a, for milling using radius programming:
N5 G03 G90 X60 Y15 U15 LF . . . i.e. tool traverses from point 1 to point 2, or conversely –
N10 G02 X45 Y30 U15 LF . . . i.e. tool traverses from point 2 to point 1.

Example 2. Relating to Fig. 1.10b, for turning using either interpolation parameters or radius programming:

interpolation parameters:
    N5 G03 G90 X40 Z80 K15 I0 LF
    N10 G02 X70 Z65 K0 I15 LF
        N5 = tool traverse from P1 to P2
        N10 = tool traverse from P2 to P1

radius programming:
    N5 G03 G90 X40 Z80 B+15 LF
    N10 G02 X70 Z65 B+15 LF
        N5 = tool traverse from P1 to P2
        N10 = tool traverse from P2 to P1

Helical Interpolation

Helical interpolation can be achieved by the simultaneous motions of three linear axes which are perpendicular to each other. Such interpolation necessitates programming a circular arc and a straight line, which is perpendicular to the end point of such an arc in a single block. When the program is processed, the individual motions of the axis slides are combined in such a manner that the tool describes a helix with a constant lead. By choosing either G16, G17, G18, or G19 the circle plane is selected. It is always necessary to program the axis coordinates X, Y and Z. The axis for the linear interpolation motion can be programmed either before or after the circular arc/circle has been defined. Whenever a fourth axis is incorporated onto the machine tool and used for the circular interpolation mode, the linear interpolation requirement is selected using a corresponding parallel axis. The programmed feedrate is adhered to on the circular path: however this is not the case for the actual tool path.

In Fig. 1.11a, a helical cutter path is required to generate this component feature, as shown in a typical program:

Helical interpolation:
    30% LF
    N1 G00 X0 Y25 Z1 S800 M3 LF
    N2 G01 Z-20 F150 LF
    N3 G02 X0 Y-25 Z-10 I J-25 LF
    *
    N5 M30 LF
        N1 = XY plane automatically selected (default condition) with tool traversing
           to P01

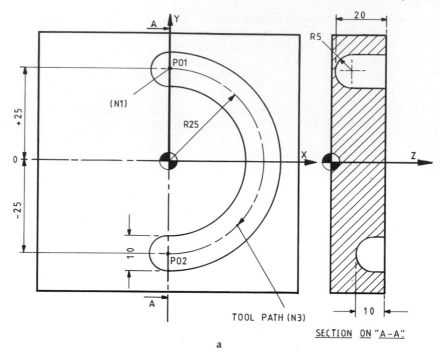

a

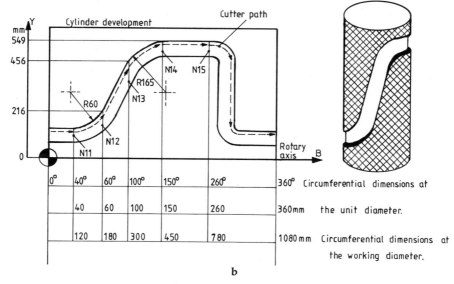

b

Fig. 1.11. a Helical interpolation. b Cylindrical interpolation. [Courtesy of Siemens.]

N2 = linear interpolation with infeed to Z-20 depth
N3 = tool clockwise traverse on a helix (G02) to P02
* N4 = rapid traverse clear – not shown
N5 = end of program

Cylindrical Interpolation

Using cylindrical interpolation allows the machining of cylindrical paths with one rotary and one linear axis, together with a constant rotary table diameter. It is also possible to program both linear and circular contours using an "intersection cutter radius compensation". In this case the position of the rotary axis is entered in degrees and then it can be converted to circumferential dimensions of the working diameter internally by the controller.

Typical details which must be considered for a specific controller are as follows:

$$\text{the ratio "P"} = \frac{\text{machining diameter}}{\text{unit diameter}}$$

and is programmed via a "G92 P" for this purpose. Input system "unit diameter" in metric units is 114.592 mm. This "unit diameter" (d) is derived from the relationship;

$$\pi \times d = 360$$

therefore the "unit diameter" (d) $= \dfrac{360}{\pi}$ (mm)

No characters other than the axis name can be written in a block containing G92 P..

Example. N..G92 P.. C LF
Where, P.. = the factor for the unit circle, working diameter/unit diameter
     C = the rotary axis

Typical input resolution for "P" is 10^{-5} and the action for the factor for the unit circle is modal, until it is reset or reprogrammed using either an M02 or M30. The programmed feedrate is maintained on the contour and if the factor is not 1, this axis – the "C" can only be interpolated with one further axis. This factor 1 must be set for any interpolation with more than two axes.

The following programmed example relates to the illustration of cylindrical interpolation given in Fig. 1.11b:

N10 G92 P3 B LF
N11 G01 G42 B40 Y200 LF
N12 G03 B60 Y216 P+60 LF
N13 G01 B100 Y456 LF
N14 G02 B150 Y549 P+165 LF
N15 G01 B260 LF
.
.

N26 G92 P1 B LF
     N10 = selection of cylindrical interpolation
     N26 = cancellation of cylindrical interpolation

Polar Coordinates

Polar coordinate programming is invariably used when holes on pitch circle diameters (PCD) are required – angular faces, or similar manufacturing activities. Normally drawings are dimensioned with an angle and a radius (see Fig. 1.4c) which can be entered directly in the program with the aid of polar coordinates. Polar coordinates can be programmed in a variety of ways using the following preparatory functions:

G10 – linear interpolation, rapid traverse
G11 – linear interpolation, feedrate (F)
G12 – circular interpolation, clockwise
G13 – circular interpolation, anti-clockwise

So that the controller can determine the cutter's traverse path, it requires the centre point, radius and angle to be known. The centre point is entered in the usual manner using the perpendicular coordinates (X, Y and Z), together with absolute position data at the initial stage of programming. A subsequent incremental position data input using G91 may be used, although this method always refers back to the last centre point programmed. The action of employing a centre point entry is modal and can be reset by means of either M02 or M30. The radius can be programmed via a "B" or "U" address, without a sign, and the angle is entered under an "A" address, also without a sign, with an input resolution of 10^{-5}. Such programming always refers to the first positive axis of the centre point coordinates – the reference axis and this positive direction corresponds to an angle of 0°.

In the following examples of programming polar features on components (Fig. 1.12) we can gain an insight into its flexibility and simplicity in producing part programs. In the first instance we will consider programming polar coordinates for holes dimensioned with respect to a common centre point (Fig. 1.12a):

N10 G10 G90 G81 X.. Y.. A.. U(r1) R1 = ..Ra = .. R11.. LF
N15 G10 A.. U(r2) LF
N20 G10 A.. U(r3) LF
N25 G10 A.. U(r4) LF
N30 G80 LF
     X.. Y.. = centre point of the polar coordinate system
     A.. U(r) = hole positions in polar coordinates

NB: G10 must be programmed in each block, since G81 is terminated with a rapid traverse.

In Fig. 1.12b, polar coordinates are used to mill a hexagon using the following word address program format:

N12 G11 G90 X50 Y35 U20 A0 L5 (POINT; P1)
N13 A60 LF (P2)
N14 A120 LF (P3)
N15 A180 LF (P4)
N16 A240 LF (P5)
N17 A300 LF (P6)
N18 A0 LF (P1)

NB: The angles are referred to the X-axis since the centre point coordinate in the X-plane is programmed first.

In the next two programmed examples using the preparatory functions G12 and G13, they affect a traversing motion along a circular arc between two points. In these

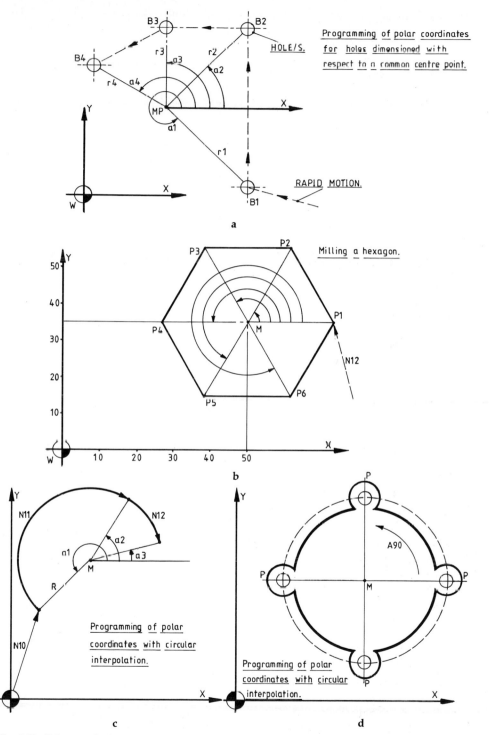

Fig. 1.12. Polar coordinate programming using **a** G10, **b** G11, **c** G12 and **d** G13. [Courtesy of Siemens.]

examples the programming of polar coordinates with circular interpolation will be considered.

Example of the preparatory function G12 (see Fig. 1.12c):

.
.
.

    N10 G11 X50 Y30 A120 U25 LF
    N11 G12 A75 LF
    N12 G12 A15 LF
    .
    .
    .

Example of the preparatory function G13 (see Fig. 1.12d):

    %40 LF
    N5 G00 G90 X120 Y100 LF
    N10 G13 X100 U20 Y20 A0 F100 LF
    N12 G81 R2=...R3=...R11=...LF
    N15 A90 LF
    N20 G81 R2=...R3=...R11=...LF
    .
    .
    .

    N35 G00 Z100 M30 LF
        N5 = hole position P1 approached
        N10 = polar coordinate selection and position
        N12 = call of bore cycle
        N15 = approach of P2
        N20 = call of bore cycle, etc.
        N35 = traverse clear

1.4.5 Feed Motions

The characteristic of the feedrate is that it determines the machining speed and is adhered to with every kind of interpolation, even when tool offsets are required to machine a contour. The value programmed under an "F" address remains in the program (i.e. a modal function) until it is superseded by a new "F-value". Any "F-value" is deleted either at the end of the program or upon reset and obviously it is a requirement to program the feed function prior to use. When a specific feedrate has been programmed it can be modified at the operator's panel – normally in the range from 1% to 120% – by means of the "feedrate override" switch. The programmed value corresponds to the 100% setting on the "feedrate override" switch.

Feedrates are programmed in either mm/min, or mm/rev, by a selection of the following "F-functions":

    G94 F.. = feedrate in mm/min
    G95 F.. = feedrate in mm/rev
    G96 F.. = feedrate in mm/rev
    G96 S.. = constant cutting speed – "S" denotes the feedrate is in m/min (see below)

Constant Cutting Speed

"Constant cutting speed" is a very useful function on a turning machine, whereby the controller determines the spindle speed for the current diameter with reference to the programmed cutting speed. The relationship between the turned diameter, its spindle speed and feed motion, permits optimum matching of the part program to the turning centre, the material being machined, and the cutting tool. The zero point of the "X-axis" must be at the centre of the turning centre and this is ensured by the reference point. When we calculate the spindle speed using the "constant cutting speed" technique, the following conditions must be considered:

actual spindle speed value

zero offset in the X-direction, using, for example, one of the following functions:
the settable zero offset (G54, G55, G56, G57)
programmable zero offset (G58, G59)
external offset

The "constant cutting speed" can be cancelled using a G97 preparatory function, with the final speed – relating to the previous G96 – being retained as the constant speed. If the X-axis is moved without further machining, an undesirable speed change can be avoided which is an aid to further programming. It is also possible to considerably reduce a set feedrate on many controllers using an M-function. The programmed feedrate can be reduced by 1:100 with an M37 and may be reset with M36, if either a superfine finish is required, or slow feeding is necessary – rather than a "dwell" – to minimise cutter vibration and its subsequent affect on the workpiece.

1.4.6 Thread Cutting – an Introduction

The machining of threads can be undertaken on both turning and machining centres and in the latter case the machine must be equipped with a helical interpolation capability – if thread milling is necessary. A large range of threads for specific applications can be cut, such as:

constant lead threads
variable lead threads
single or multiple-start threads
tapered threads
external or internal threads
transversal threads

Before we consider some examples of the part programming such threads as described above, it is worth discussing the variety of techniques that is available for producing infeeds when cutting V-form threads on turning centres.

Thread Infeed Techniques

There are a range of forming and partial forming/generating methods for producing V-form threads; some of the more popular methods are illustrated in Fig. 1.13.

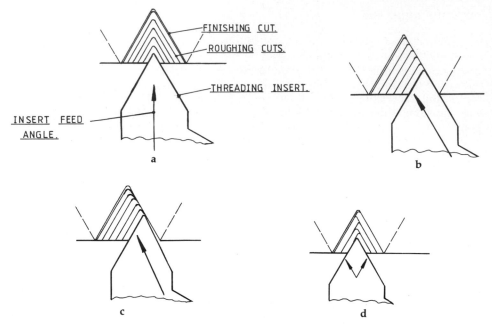

Fig. 1.13. Various methods of producing thread infeeds. **a** Radial infeed (plunge). Metal is removed on both sides of the insert simultaneously. **b** Flank infeed (half-angle). This gives a more easily formed chip and better heat dissipation. **c** Modified flank infeed. There is less wear of the trailing edge and a better surface finish on the corresponding flank. **d** Alternating flank infeed. Both edges are fully utilised, which means a longer insert life.

Radial Infeed

The first example we will consider is that shown in Fig. 1.13a which is termed the radial, or plunge infeed technique. This is the method which is usually adopted when cutting square threads, but can be used to machine V-form threads. It is purely a forming method, which means that the profile produced on the threading insert is replicated on the workpiece – hence the term "forming". However, it suffers from a serious disadvantage when used to form threads – as this method is not normally associated with "canned cycles" – which means that as the full depth of cut is approached (after successive passes) cutting occurs along the whole insert profile making a torn, or poorly finished thread very likely. Such a thread condition is a function of a range of tool, work and cutting data factors:

tool geometry – promoting a poor chip-breaking tendency, together with long cutting edges

workpiece ductility/affinity to the tool, causing a likelihood of torn threads at full depth

feeds/speeds/infeeds progression being inappropriate to good chip-breaking ability

angular relationship of the tool to the workpiece – whether the form is "square" and "true" to the part

Flank Infeed

The flank or alternatively, as it is often known, the half-angle infeed technique is depicted in Fig. 1.13b, which is the popular method used for screwcutting operations on conventional lathes. It is a combination of generating and forming, with the generated flank being the product of successive infeeding passes down this flank, whereas the adjacent flank is formed at full thread depth by the threading insert's profile. Chip forming is more efficient in this case than for the previously mentioned radial infeed technique, but flank wear is associated with the forming insert's clearance flank in particular. This will limit the number of threading passes and hence the screwthreads produced, as this greater flank wearing tendency on the generating edge affects the finish on this flank of the thread. "Canned cycles" sometimes use this infeed technique for screwthread production.

Modified Flank Infeed

The modified flank infeed technique (Fig. 1.13c) answers most of the criticisms levelled at flank infeeding operations. It is a curious mixture of partial generation and forming in the main, and the final pass is at full forming to depth along the thread. As its name implies, the half angle is modified to a more acute angle – usually between 27° and 29° – which can be adjusted accordingly. This means that as the so-called generating infeed is 1°–3° less than the desired angle there is less wear on the tool's trailing flank as a result. When full depth is reached, the complete profile of the thread is reproduced by the threading insert at this last pass. "Canned cycles" often use this infeed technique as it reduces tool wear and improves thread quality over the previous techniques mentioned.

Alternating Flank Infeed

Probably the most satisfactory thread production method is that of the alternating flank infeed technique (Fig. 1.13d), from the point of "balanced" wear on the threading insert. As its name implies, with consecutive threading passes the bias of the insert's cutting flank changes, so that on one pass the left-hand flank cuts and on the next pass the right-hand flank cuts. This technique, when incorporated into a "canned cycle" for threading, fully utilises each cutting edge, which minimises flank wear whilst increasing tool life, and part quality is enhanced further as a result; particularly when cutting with "full depth" threading inserts which allows a limited amount of stock diameter to be removed on the final threading pass, thereby ensuring a true thread profile.

Not all of these infeed techniques are available on every CNC controller and, in fact, other techniques not discussed here may be available, but they will all – to a greater or lesser extent – cut a thread satisfactorily. In the remaining comments in this section we will consider some of the preparatory functions used in part programming threads and go on to mention certain programming considerations required, if a satisfactory threading operation is to be undertaken.

The following preparatory functions are available for machining threads:

G33 – thread cutting with a constant lead
G34 – thread cutting with a linear lead increase
G35 – thread cutting with a linear lead decrease

The usual method of entering thread data is in terms of thread length and lead, as the following examples will show. Thread length is entered under the corresponding path address, with the start–stop and overrun zones at which the feedrate is increased or decreased being taken into consideration. The numerical values can be entered using either absolute, or incremental position data. Thread lead is entered via such addresses as "I, J and K". For longitudinal turned threads the lead is entered under a "K", with transversal threads it is entered using "I", whereas tapered threads are entered using a "K" address. These addresses must always be entered using incremental position data, but without a sign. Most controllers have a standard input resolution for the thread lead of 0.001 mm revolution, with the programming of the lead of 0.001 mm– 2000 mm. Typically, if a thread lead of 1 mm is programmed as the input resolution, it is possible to obtain a resolution of 0.01 mm/revolution with the M37 function. Right- and left-hand threads are programmed using the spindle direction of rotation functions M03 and M04 and these, together with the speed, must be programmed in the block prior to an actual thread cutting operation. This presetting of such functions is necessary in order to permit the spindle to run-up to its nominal programmed speed, as the following example shows:

    N10 S500 M03 LF
    N15 G33 Z... K... LF

This programming logic means that the feed does not begin until the zero mark is reached on the pulse encoder, in order to permit threads to be cut in several passes, thus ensuring that the threading insert will always enter the workpiece at the same point on the circumference, with the cuts being implemented at the same cutting speed and, as such, preventing discrepancies in "following errors".

The "feedrate override switch", "feed-off key", "spindle speed override switch" and the "single block mode" have no effect when cutting the thread. Furthermore, the feedrate programmed using the "F-word" will remain stored and only becomes effective again when the next G01 function is programmed.

In this section, mention was made of "canned cycles". The following section explains, using practical programming examples, how such "cycles" can reduce the part program's length and minimise the effort required by the programmer. We will also gain an appreciation of how such "canned cycles" are engaged for specific cutting operations for particular types of threads and the relationship of the threading insert to the workpiece during consecutive threading passes.

1.4.7 Programming Threads on Turning Centres

The following part programming examples illustrate solutions for a variety of thread forms.

Thread with Constant Lead (Fig. 1.14a)

NB: The feedrate preparatory function "F" is not programmed here, since the feedrate is linked directly to the spindle speed via a pulse encoder.

Example 1. Absolute position data input :

    N20 G90 S... LF
    N21 G00 X46 Z78 LF   (P1)

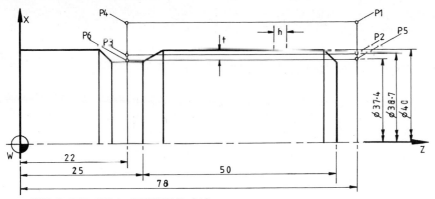

THREAD DATA FOR A CYLINDRICAL BAR;

Lead h = 2 mm, Thread depth t = 1·3 mm, Radial infeed direction.

a

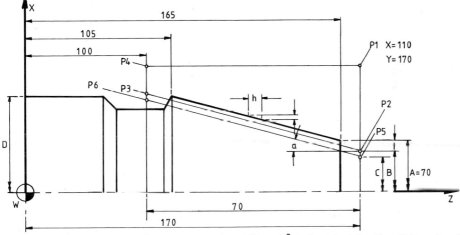

THREAD DATA; Lead h = 5 mm, Thread depth t = 1·73, a = 15°, Radial infeed direction. Both end position coordinates must be written. The lead "h" is entered under "K".

CALCULATION OF THE THREAD START & END POSITION COORDINATES;

(A,B,C & D are diameters). 1st cut P2...P3, t = 1 mm.

2nd cut P5...P6, t = 1·73 mm.

$A = 70$,

$B = A - 2t = 66·54$,

$C = B - 2 (5 * \tan a) = 63·86$,

$D = C + 2 (70 * \tan a) = 101·366$,

$K2 = h = 5$,

$1 = h * \tan a = 1·34$ mm.

NB $*$ = Multiplication character.

b

Fig. 1.14. a Thread with a constant lead. **b** Thread on a tapered bar. [Courtesy of Siemens.]

```
N22 X38.7 LF                    (P2)
N23 G33 Z22 K2 LF               (P3)
N24 G00 X46 LF                  (P4)
N25 Z78 LF                      (P1)
N26 X37.4 LF                    (P5)
N27 G33 Z22 K2 LF               (P6)
N28 G00 X46 LF                  (P4)
```

Example 2. Incremental position data input:

```
N20 G91 S.. LF
N21 G00 X-... Z-... LF          (P1)
N22 X-3.65 LF                   (P2)
N23 G33 Z-56 K2 LF              (P3)
N24 G00 X3.65 LF                (P4)
N25 Z56 LF                      (P1)
N26 X-4.3 LF                    (P5)
N27 G33 Z-56 K2 LF              (P6)
N28 G00 X4.3 LF                 (P4)
```

Thread on a Tapered Bar (Fig. 1.14b)

Absolute position data input:

```
N31 G90 S... LF
N32 G00 X110 Z170 LF                (P1)
N33 X65.86 LF                       (P2)
N34 G33 X103.366 Z100 K5 LF         (P3)
N35 G00 X110 LF                     (P5)
N36 Z150 LF                         (P1)
N37 X63.86 LF                       (P5)
N38 G33 X101.366 Z100 K5 LF         (P6)
N39 G00 X110 LF                     (P4)
      Calculation of points "P2" & "P3":
      X(P2) = C + 2mm = 65.86mm
      X(P3) = D + 2mm = 103.366mm
```

Transversal Thread (Fig. 1.15a)

Thread details: Lead h = 2mm
Thread depth t = 1.3mm
Infeed direction: perpendicular to the cutting direction
Absolute position data input:

```
N41 G90 S... LF
N42 G00 X4 Z82 LF       (P1)
N43 Z79.35 LF           (P2)
N44 G33 X36 I2 LF       (P3)
N45 G00 Z82 LF          (P4)
N46 X4 LF               (P1)
N47 Z78.7 LF            (P5)
```

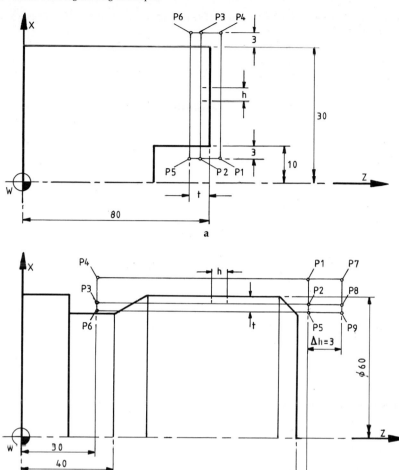

Fig. 1.15. a Cutting a transversal thread. **b** Cutting a multiple thread. [Courtesy of Siemens.]

```
N48 G33 X36 I2 LF      (P6)
N49 G00 Z86 LF         (P4)
N50
.
.
.
```

Multiple Threads (Fig. 1.15b)

Any thread cutting operation always begins at the synchronisation point of the zero mark of the pulse encoder. The feed will not be enabled unless a signal is received from the digital rotary transducer. The starting point for thread cutting can be offset in the program, making it possible to cut multiple threads. A "start" of a multiple thread is programmed in the same way as a "single-start" thread. When the first "start" has been completely machined, the start position in this case is offset using an (h) character, allowing the next "start" to be machined and is calculated in the following manner:

$$h = \frac{\text{thread lead}}{\text{number of starts}}$$

The various "starts" must be executed at the same spindle speed, in order to avoid discrepancies in the following error.

Thread details: Lead h = 6 mm
Thread depth t = 3.9 mm
Number of starts = 2

In this example (Fig. 1.15b), each "start" is machined in two steps. When the first "start" has been fully machined, the second "start" is machined by offsetting the start position by:

$$h' = \frac{\text{thread lead}}{\text{number of starts}} = \frac{6}{2} = 3\,\text{mm}$$

Absolute position data input:

```
N35 G90 S... LF
N36 G00 X66 Z115 LF    (P1)
N37 X56 LF             (P2)
N38 G33 Z30 K6 LF      (P3)
N39 G00 X66 LF         (P4)
N40 Z115 LF            (P1)
N41 X52.2 LF           (P5)
N42 G33 Z30 K5 LF      (P6)
N43 G00 X66 LF         (P4)
N44 Z118 LF            (P7)
N45 X56 LF             (P8)
N46 G33 Z30 K6 LF      (P3)
N47 G00 X66 LF         (P4)
N48 Z118 LF            (P7)
N49 X52.2 LF           (P9)
N50 G33 Z30 K6 LF      (P6)
N51 G00 X66 LF         (P4)
```

This completes the review of thread cutting applications although it is by no means an exhaustive account of the thread programming solutions. Notable omissions include: variable lead threads and worms, with thread forms such as acme, buttress and so on, not being described. However, the reader should by now have gleaned a reasonable understanding of thread production techniques on turning centres.

1.4.8 Miscellaneous Functions

The miscellaneous functions contain the primary technical specifications which are not programmed in the words provided with the address letters "F, S and T", where such functions are:

miscellaneous functions "M"
spindle speed "S"
tool number "T"
auxiliary function "H"

Up to three "M-functions" can typically be contained in any block, together with an "S-, T- and H-function". Such functions are output to the interface controller, usually in a specific sequence as follows: "M, S, T, and H". Whether these functions are output before or during the axis movement is specified by the machine data. When functions are output during the axis movement, the following details apply: if a new value must be active before an axis is traversed, the new function must be written in the preceding block.

Several miscellaneous functions are defined in the standard (DIN 66025 part 2), whilst others are defined by the machine tool manufacturer. The following "M-functions" are in general use on the majority of controllers:

M00 programmed stop (unconditional)
This function permits interruption to the program – perhaps to perform a measurement. On completion of the measurement the machine tool can be restarted, with the information entered being retained. This M00 function is active during all automatic operating modes – whether or not the spindle drive is also stopped must be determined from the specific programming instructions for each machine. Furthermore, an M00 function is also active in a block without position data.

M01 programmed stop (conditional)
The M01 function is similar to an M00, but it differs in that it is only active when the "conditional stop active" function has been activated, usually using a "soft key".

M02 programmed
This function signals the end of the program and resets the program to the beginning of the first block. It is always written in the last block of the program and the controller is reset. M02 can be written in a separate block, or alternatively, in a block containing other functions.

M17 subroutine end
Such a function is always written in the last block of a subroutine and can be incorporated with other functions, but cannot be present in the same block as a "nested subroutine".

M30 program end
This function fulfils the same as an M02.

M03, M04, M05 and M19 main spindle control
NB: M19 is only used with a pulse encoder in the main spindle. When an analog spindle speed version is present, the output uses the following M-words for spindle control:

   M03, clockwise spindle direction of rotation
   M04, anti-clockwise spindle direction of rotation
   M05, non-orientated spindle
   M19, orientated spindle

Often, an extended address notation with an output of the spindle number – dependent upon the controller type – may be used for miscellaneous functions in this group as the following example shows:

M2 = 19 S..
where: 2 = spindle number
       19 = M-function 19
       S.. = angle under S

"M19S.." can be used to perform an orientation and main spindle stop, with the relevant angle being programmed via an "S" in degrees. This angle is measured from the zero mark in a clockwise rotation and is a modal address. This modal function value is stored under the "S", it being valid for any angle, which may be used as a repeated stop using simply M19. It should be said that the M19 function does not cancel either M03 or M04.

Freely assignable miscellaneous functions

All miscellaneous functions – the exceptions being M00, M01, M02, M03, M04, M19, M30, M36, M37 – are freely assignable. An extended notation must always be used and the channel number specified for all freely assignable miscellaneous functions, typically as shown below:

M3 = 124
where: 3 = CNC channel number
     124 = M-function 124

NB: Further details on the use of these and other functions can be obtained in the specific program key. The meaning of such functions is defined in DIN 66025.

Spindle Function "S"

The data listed below can be entered as an option via the address "S":

spindle speed in rev/min, or 0.1/min *
cutting speed in m/min, or 0.1/min *
spindle speed limitation in m/min, or 0.1/min *
spindle dwell in revolutions **
*: the speed and cutting speed must be programmed in the same input format.
**: dependent on the type of controller.

Further, an extended address notation must always be used and the spindle number specified for an "S-word" as the following example shows:

S2 = 1000
where: 2 = spindle number,
     1000 = spindle speed.

Auxiliary Function "H"

Normally, only one auxiliary function per block can be entered under the "H" address for machine switching functions, or movements not covered by numerical control. Depending on the control system used, "H" can be programmed with up to 4–8 decades and such function meanings are described in programming instructions of the machine tool manufacturer.

The extended address notation must be used and the channel number specified for the "H-word" as described below:

H.. - ...

Tool Number "T"

The tool number determines the specific tool required for a given machining operation, as the following example shows:

T1234...
where: 1234... = tool number, the maximum is 4 or 8 decades depending on the system.

The extended address notation must be used and the channel number specified for the "T-word", such as:

T.. = ...

This completes the description of the relative merits of miscellaneous functions and in the next section we will consider the structure of sub-routines and "nesting" together with how they are used in CNC applications. Later, subroutines will be used in conjunction with parametric programming techniques.

1.4.9 Subroutines

Whenever the same machining operation must be performed repeatedly on a workpiece, it can be entered as a subroutine and "called" as often as desired within the part program, or activated using "manual data input" (MDI). A typical standard memory in a controller will be able to store up to 200 average length part programs and subroutines jointly, which can be extended still further if program memory is expanded.

Normally, it is preferable to write subroutines using incremental position data, with the tool being set to the position in the part program just prior to the start of a subroutine call. The workpiece machining sequence, with the tool positioned correctly, can be used to repeat the operation at various points on the workpiece without the need to modify the dimensions in the subroutine.

The subroutine structure (Fig. 1.16a) comprises the:

subroutine start (header)
subroutine blocks
end of subroutine

The subroutine start is typically comprised of the address "L" and the 3 or 4 digit subroutine number.

NB: The subroutine start is not in tape format. When the subroutine end is called, this is used to return to the part program and is defined by the M17 end character. An M17 function is written in the last block of the subroutine and it is permissible to write other functions (except the address "L") in this block.

The subroutine can be called in a part program via the address "L" with the subroutine number and the execution of the repetitions with the "P" address, as the following example illustrates:

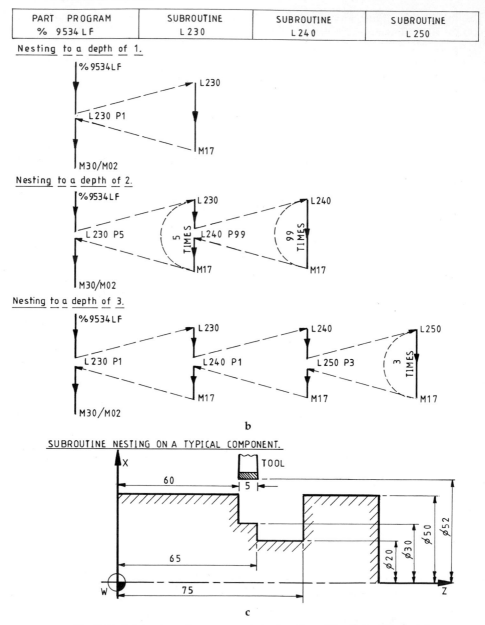

Fig. 1.16. Substructure of program nested subroutines. [Courtesy of Siemens.]

L123 P1
where: L123 = subroutine (1 . . . 999)
P1 = number of repetitions

The following logic should be noted during programming using subroutines:

the subroutine call must not be written in a block together with M02, M30 or M17

if the subroutine is called whilst the cutter radius compensation (CRC) function is active, this can create some problems, therefore see the section on p. 68 on special cases for CRC – blocks without path addresses should be referred to for an understanding of tool path motions

when a subroutine call is written in a block that contains other functions, the subroutine is called at the end of the block

Subroutine Nesting

It is possible not only to call subroutines from the part program, but also from other subroutines and this process is called "subroutine nesting" (Fig. 1.16b). Normally it is only possible to "nest" a series of subroutines to a finite depth, which typically might be three or four – depending upon the controller, with one particular controller having the ability to nest to a depth of 28 subroutines. In Fig. 1.16b, a typical programming application for the nested subroutines to a depth of three, might be:

first nested subroutine – to drill a hole pattern using "canned cycles", such as those shown in Fig. 1.30a,b for either a linear drilling pattern (G26), or a circular drilling pattern (G28) respectively

second nested subroutine – within the first nested subroutine another "canned cycle" can be called to drill the hole pattern chosen, such as the standard drilling cycle (G81) shown in Fig. 1.28a

third nested subroutine – finally, all, or some of these drilled holes might require tapping and this allows us to call up the final canned cycle for tapped holes (G84) shown in Fig. 1.29a, before returning to the main program

The use of such nesting and canned cycle programming activities has the benefit of considerably reducing the number of blocks of information necessary to complete the part program. This not only simplifies such programming activities, but utilises less memory capacity, giving the additional benefit of enabling one to store many more programs in the controller's memory.

As a practical example of the ability of nested subroutines to shorten programs and at the same time give the reader an understanding of the programming logic, the following turning example based upon Fig. 1.16c has been produced:

```
%9534 LF
N1 G90 G94 F.. S.. D.. T.. M.. LF
N2 G00 X52 Z60 LF
N3 L230 P1 LF          (subroutine call)
 .
 .
 .
N90 M30                (end of part program)
          (subroutine structure)
L230 LF                (start of subroutine)
NG G91 G01 X-11 LF
```

```
N2 G09 X11 LF
N3 L240 P2 LF                (nested subroutine call)
N4 M17 LF                    (end of subroutine – back to part program)
      (nested subroutine structure)
L240 LF                      (start of nested subroutine)
N1 G91 G00 Z5 LF
N2 G01 G09 X-16 LF
N3 G00 X16 LF
N4 M17 LF                    (end of nested subroutine – back to subroutine)
```

1.4.10 Parametric Programming

Parametric programs offer the programmer a flexible, concise and powerful programming aid and they are used in a program to represent a numerical value of an address. Parameters are assigned values within the program and as such can be used to adapt programs to several similar applications; for example:

different feedrates *

different spindle speeds *

differing operating cycles

components with varying aspect ratios – such as those used in a "Group Technology" (GT) approach to component manufacture (see Fig. 1.17a)

calculating mathematical expressions for – trigonometric functions, addition, subtraction, multiplication and division of numerical values in specific blocks

* useful when modifying different materials to be machined, without changing the main cutting data factors in the part program which might otherwise be necessary when programming using the "traditional" approach.

Often such parametric programming adaptability is termed "free-variable parameters" and a typical assignable parameter might comprise the address "R" and a number with up to three digits. Typically up to three hundred parameters are selectable in the basic configuration within many controllers. Such parameters can be sub-divided into:

transfer parameters

computing parameters

channel-dependent/independent parameters **

central parameters **

** these parameters are explained in manufacturers' manuals and are outside the scope of this section on parameter programming.

Parameter Definition

Quite simply, a parameter definition may be used to assign certain numerical values together with signs to the various parameters. Such parameters can be defined either in part programs, or in subroutines. A typical parameter definition might be:

R1 = 10 LF

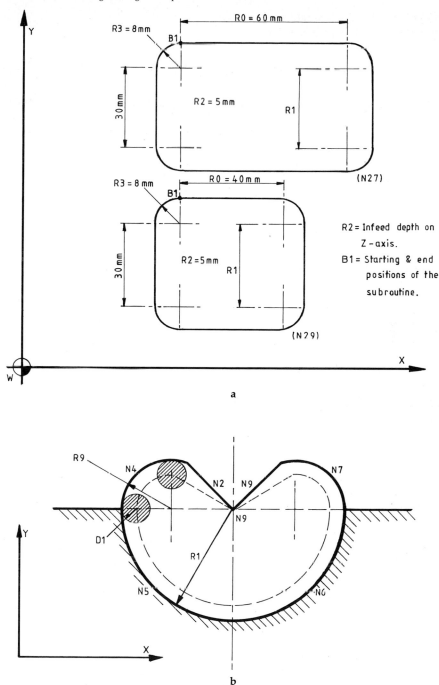

Fig. 1.17. Parameter programming for similar workpiece geometries and complex shapes. **a** Milling rectangles with variable aspect ratios in the X–Y plane. **b** Using "parametrics" to machine an internal semicircle. [Courtesy of Siemens.]

In a single block, it is acceptable to have the parameter definition, a subroutine call, together with the switching functions present. The value defined for a parameter is assigned directly to the address, as the following example illustrates:

%5772
N1...
.
.
.
N37 R1=10 R29=20.05 R5=50 LF
N38 L51 P2 LF   (subroutine call)
M39 M02 LF   (end of program)
L51
N1 Z=−R5 B=−R1 LF
N2 X=−R29 LF
.
.
.
N50 M17 LF   (end of subroutine, back to part program)
        where N37 = parameter definitions
                  N38 = subroutine call "51", with 2 repetitions

Parameter Calculations

The linking of parameters. As we have already mentioned, all four basic arithmetic operations may be used when parametric programming. It is important to the result of any calculation to link the parameters in a specific sequence as the following tabulation shows:

a6.5	Arithmetic operation	Programmed execution/ Arithmetic operation
Definition		$R1 = 100$
Assignment		$R1 = R2$
Negation		$R1 = -R2$
Addition		$R1 = R2 + R3$
Subtraction		$R1 = R2 - R3$
Multiplication		$R1 = R2 * R3$
Division		$R1 = R2/R3$

When the result of an arithmetic operation is written in the first parameter of a link, its initial value will be overwritten and as such is lost upon linking. However, the values of any second and/or third parameters are retained. If the value of one parameter is to be assigned to another, the following logic is valid:

R1 = R3 LF, as we can see in the tabulation above.

Calculations Using Numbers and Parameters

(i) *The addition and subtraction of numbers and parameters*

With any parameter it is possible to add to the value of an address, or to subtract from it accordingly. The sequence which follows must be used in such cases: address, numerical value, parameter. When no sign is specified, it is assumed that a positive sign (+) will be the default. In the example below we can see how the parametric logic is used to determine specific numerical values:

N38 R1=9.7 R2=−2.1 LF
N40 X=20.3+R1
N41 Y=32.9 R2
N42 Z=19.7−R1

The numerical result of these calculations:

(Line 40) X = 30
(Line 41) Y = 35
(Line 42) Z = 10

(ii) *Calculations using numbers and parameters*

Unfortunately it is not possible to multiply, divide, add, or indeed subtract absolute numbers and "R-parameters". Therefore under these circumstances we must use "auxiliary parameters", as described below:

Not permissible would be: R10=15+R11
permissible calculations are: X=10+R11

Let us now look at a practical example of such a technique. Assume that the parameter R2 must be divided by 2:

R3=2 Definition of an auxiliary parameter, hence:
R1=R2/R3.

The result of the calculation is contained in R1, with the values of R2 and R3, the auxiliary parameters, being retained.

Parameter String

The following example illustrates a typical parameter string:

R1=R2+R3−R4*R5/R6 . . . . R10

As we can appreciate from this expression, all the four basic arithmetic operations are permissible in any sequence. It is acceptable to link up to ten parameters together in a parameter string and such calculations are performed as follows (based upon the example shown above):

Step 1: R1=R2+R3
Step 2: R1=R1−R4
Step 3: R1=R1*R5
Step 4: R1=R1/R6

.
.
.

i.e. Step 1   R1 = R2 + R3
                    ↓
      Step 2   R1 − R4
                    ↓
        Step 3   R1 * R5
                      ↓
          Step 4   R1 / R6
                        ↓
                       R1

NB: It is acceptable to perform any number of arithmetic operations in a block, typically multiplication, parameter strings, addition, and so on, with the limitation being the maximum permissible block length of 120 characters in most controllers. The individual links are calculated by the controller in the programmed sequence.

Such calculations are all very well, so let us look at how parametric programming can be put into practice and the examples chosen for the reader to gain a more complete appreciation are typical milling operations.

Milling Rectangles Using Parametric Programming

The example chosen (Fig. 1.17a) illustrates the variable aspect ratios of rectangles which require milling. The subroutine written below permits a rectangle whose sides are parallel to the machine axis to be machined in the "X–Y" plane:

Example 1 (Fig. 1.17a)

```
L46
N5 G01 G64 G91 Z=−R2 LF
N10 X=R0 LF
N15 G02 X=R3 Y=−R3 I0 J=−R3 LF
N20 G01 Y=−R1 LF
N25 G02 X=−R3 Z=−R3 I=−R3 J0 LF
N30 G01 X=−R0 LF
N35 G02 X=−R3 Y=−R3 I0 J=R3 LF
N40 G01 Y=R1 LF
N45 G02 X=R3 Y=R3 I=R3 J0 LF
N50 G01 Z=R2 LF
N55 M17 LF

     Subroutine call:
N26 G90 X... Y... LF
N27 L46 P1 R0=60 R1=30 R2=5 R3=8 LF
N28 G90 X... Y... LF
N29 L46 P1 R0=40 LF
```
        where: N26 = first starting position of the current program
               N28 = second starting position.

Summarising: by changing the "free-variable" parameters in lines and using subroutines, the variable aspect ratios of the two (i.e. large and small) rectangles can be milled respectively.

Milling an Internal Semi-Circle

In the second example (Fig. 1.17b), the subroutine shown below can be used to rough and finish mill a semi-circular profile. The contour radius and the approach and retract radii (i.e. "scroll-in and-out"), can be varied by using parametric programming. The difference between the workpiece's actual size and its design size can be checked after each cutter pass. This difference may then be entered into the program as the "additive" tool wear.

Example 2 (Fig. 1.17b)

    Subroutine.
    L11
    N1 R1=R1−R9 LF
    N2 G00 G64 G91 G17 G41 D01 LF
    N3 R1=R1+R9
    N4 G03 X=−R9 Y=−R9 I0 J=−R9 LF
    N5 X=R1 Y=−R1 I=R1 J0 LF
    N6 X=R1 Y=R1 I0 J=R1 LF
    N7 X=−R9 Y=R9 I=−R9 J0 LF
    N8 R1=R1−R9 LF
    N9 G00 G40 X=−R1 Y=−R9 LF
    N10 R1=R1+R9 M01
    N11 M17 LF

        where: N1 = Calculation of approach circle
                    N2 = Approach circle positioning
                    N3 = Working back "R1" to the original value
                    N4 = Contour approach (i.e. "scroll-in")
                    N5 = Machining
                    N6 = Machining
                    N7 = Retract from contour (i.e. "scroll-out")
                    N8 = Calculation of workpiece centre point
                    N9 = Positioning
                    N10 = Working back "R1" to the original value
                    N11 = Subroutine end.

    Subroutine call.
        %5873 LF
        N1.......
        N2 L11 P1 R1=50 R9=10 LF
        N3....... LF

1.4.11 Conversational/Blueprint/Shop-floor Programming

In order to build up a contoured profile to be either turned or milled, multi-point cycles for direct programming in accordance with the workpiece drawing ("blueprints") are provided for conversational programs. The points of intersection of the straight lines of the contour are entered as coordinate values, or alternatively, via angles. These straight lines can be joined together either directly, in the form of a corner, or rounded via radii, but chamfers also can be accommodated. Chamfer and transition radii are specified only by means of their size, with the geometric calculation being performed by the controller. The end position coordinates may be programmed using either absolute or incremental position data.

Contouring Cycle

As we can see from Fig. 1.18a,b, a contoured profile consists of the systematic build-up and assembly of discrete contoured elements (Fig. 1.18b). Later in this section the programming logic for the milled contour will be considered together with some turned examples of "blueprint" programming. Prior to that, it is important to under-

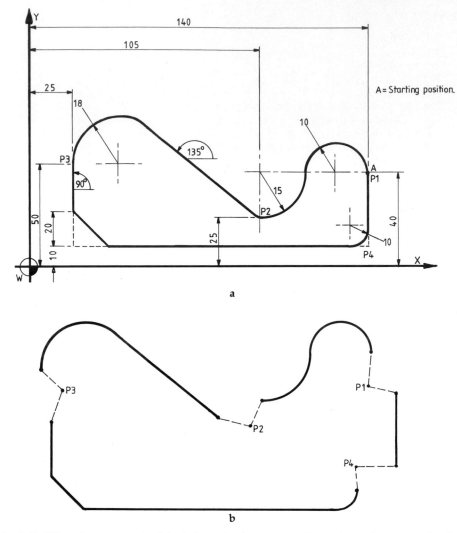

Fig. 1.18. ''Blueprint programming'' for 2-dimensional contours. **a** Contouring cycle programming for a machining centre. **b** The elements in the construction of a contouring cycle program. [Courtesy of Siemens.]

stand the use of the preparatory function G09 and its operation. Also discussed will be ''F, S, T, H, and M'' in the contouring cycle.

If the G09 function is programmed in the contouring cycle block, it will not be active until the end of the block – in other words, until the end position is reached. The G09 preparatory function is automatically generated by the controller when irregular points occur, typically corners, edges, etc., within the contouring cycle.

Linking of Blocks

The linking of blocks is possible with/without either angle inputs, inserted radii, or chamfers, in any sequence. In the following example of a profile milling operation (Fig. 1.18a), the following contouring cycles are used – circular arc – circular arc – straight line – circular arc, 3-point cycle + chamfer + radius, as shown below:

```
L168
N1 G90 G03 I-10 J0 I0 J15 X105 Y25 LF (P2)
N2 G03 A135 U18 X25 Y50 LF (P3)
N3 G01 A90 A0 X140 Y10 U-20 U10 LF (P4)
N4 Y40 LF (P1)
N5 M17 LF
```

The following programmed examples for turned parts illustrate a range of contours for either external or internal features (Fig. 1.19). In the first example (Fig. 1.19a) of external machining, the angle "a" refers to the starting position, whilst angle "b" is associated with the missing vertex. The end position can be programmed using either absolute position data G90, or incremental position data G91. Both end position coordinates must be specified. The controller determines the vertex from the known starting position, together with the two angles and the end position:

```
N10 G00 G90 X30 Z105 LF
N11 G01 A170 A135 X100 Z20 F.. LF
```

The internal machining example (Figs. 1.19b,c) can be determined from the drawing dimensions (Fig. 1.19c), with the starting position being defined anywhere outside the inner cone. The perpendiculars through the starting position and the extension of the internal cone will yield the point of the intersection "A" with the part program continuing, as follows:

```
.
N13 G00 Xstart Zstart LF
N14 G01 A90 A184 X...Z...LF
```

Finally the last contouring cycle program on a turning centre using blueprint programming, is illustrated below and is based upon the dimensional features shown in Fig. 1.19d:

```
L105
N5 G00 G90 X0 Z332 LF
N10 G01 G09 A90 X66 B-8 F0.2 LF **
N15 A180 A90 X116 Z246 B8 LF
N20 G03 A90 X116 Z246 B8 LF
N20 G03 B40 A175 X140 Z130 LF
N25 G01 A135 A180 X220 Z0 LF
N30 M17 LF
        where ** = linking with B
```

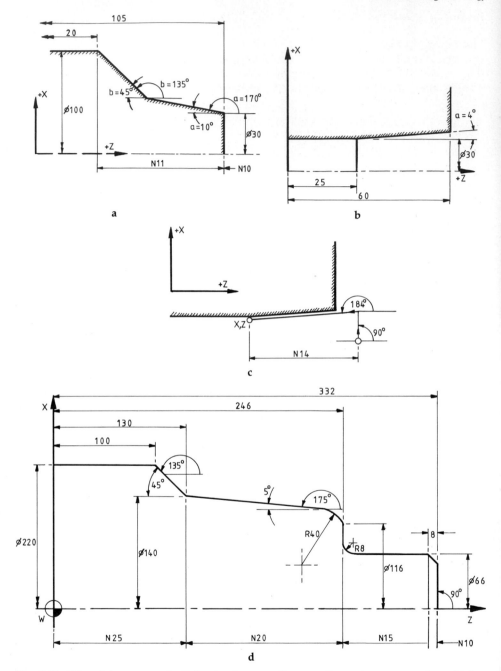

Fig. 1.19. "Blueprint programming" for turning operations. **a** Contouring – external machining. **b** Contouring – internal machining. **c** Drawing dimensions. **d** Contour cycle programming for a turning centre. [Courtesy of Siemens.]

1.4.12 The Structure of Tool Offsets and their Compensations

Tool Offset Number

In most CNC controllers the geometrical tool data for the tooling is usually stored
under the tool offset number "D", where the:
  Length ± 999.999 mm
  Radius ± 999.999 mm
Depending on the controller type, the "T-number" is usually 4–8 decades. Typically
the controller has its tool offset block sub-divided into 8 columns: P0–P7. The format
of the tool offset block may be identified by the tool type (P1) (see Fig. 1.20). On one
popular controller, either 99 or 128 tool offset blocks are available to the user, or a
number may be selected via the machine data. The tool offset can be called up in up to
three decades, via D1 to D... addresses, with its cancellation being with a D0. The
tool offset is not rescinded, however, until the corresponding axis is programmed.
    The tool number, type, geometry and the wear of all active tooling are stored in the
tool offset area of the controller. The tool geometry together with its associated wear is
updated by, for example, the measuring cycles in the CNC. Such wear when it occurs,
is summated in the controller in accordance with the machine data.

Tooling Classification

The following example shows the breakdown of the tool type P1:

  Unqualified: Type 0   Tool not defined
                   Type 1 . . . 9 Lathe tools
  Semi-qualified: Type 10 . . . 19 Tools with active length
                              compensation only (e.g. drills)
  Qualified: Type 21 . . . 29 Tools with radius and length
                       compensation (e.g. cutters)
  Qualified: Type 30 . . . 39 Tools with radius compensation and two
                       length compensations (e.g. angle cutter)

Tool Offsets on Turning Centres without Using Tool Nose Compensation

The tool offset is effectively derived from the sum of the tool length compensation
together with any external, additive compensations. Hence, the sum of such compensations would correspond with the dimension "XSF" or "ZSF" as depicted in
Fig. 1.21a, where:

                      P = theoretical tool tip
                      S = tool nose centre
                      F = slide reference point

In Fig. 1.21a, the tool's path is determined from the programmed tool nose centre "S"
and this type of length compensation is generally referred to as the "tool nose centre".
Incorporating tool compensation into the "tool nose centre" causes differences
between the old and new offset values if the tool offset number is changed. This poses
the question as to whether the difference is traversed directly following the change, or
not taken into consideration until the programmed traverse of the corresponding axis
is resolved on start-up.

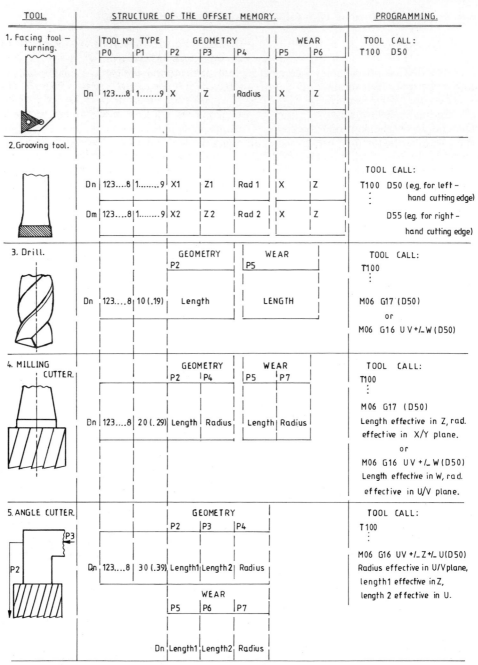

TOOL.	STRUCTURE OF THE OFFSET MEMORY.	PROGRAMMING.

1. Facing tool – turning.

	TOOL N° P0	TYPE P1	GEOMETRY P2	P3	P4	WEAR P5	P6
Dn	123....8	1.......9	X	Z	Radius	X	Z

TOOL CALL:
T100 D50

2. Grooving tool.

	P0	P1	P2	P3	P4	P5	P6
Dn	123....8	1.......9	X1	Z1	Rad 1	X	Z
Dm	123....8	1.......9	X2	Z2	Rad 2	X	Z

TOOL CALL:
T100 D50 (e.g. for left –
⋮   hand cutting edge)

D55 (e.g. for right –
       hand cutting edge)

3. Drill.

			GEOMETRY P2	WEAR P5
Dn	123....8	10 (.19)	Length	LENGTH

TOOL CALL:
T100
⋮

M06 G17 (D50)
or
M06 G16 U V +/_ W (D50)

4. MILLING CUTTER.

			GEOMETRY P2	P4	WEAR P5	P7
Dn	123....8	20 (.29)	Length	Radius	Length	Radius

TOOL CALL:
T100
⋮

M06 G17 (D50)
Length effective in Z, rad.
effective in X/Y plane.
or
M06 G16 UV +/_ W (D50)
Length effective in W, rad.
effective in U/V plane.

5. ANGLE CUTTER.

			GEOMETRY P2	P3	P4
Dn	123....8	30 (.39)	Length1	Length2	Radius

	WEAR P5	P6	P7
Dn	Length1	Length2	Radius

TOOL CALL:
T100
⋮

M06 G16 UV +/_ Z +/_ U(D50)
Radius effective in U/Vplane,
length1 effective in Z,
length 2 effective in U.

N.B. G16 is purely a setting function; no traversing is possible in these blocks. D is not coupled to plane selection.

Fig. 1.20. The structure of the tool offset memory. [Courtesy of Siemens.]

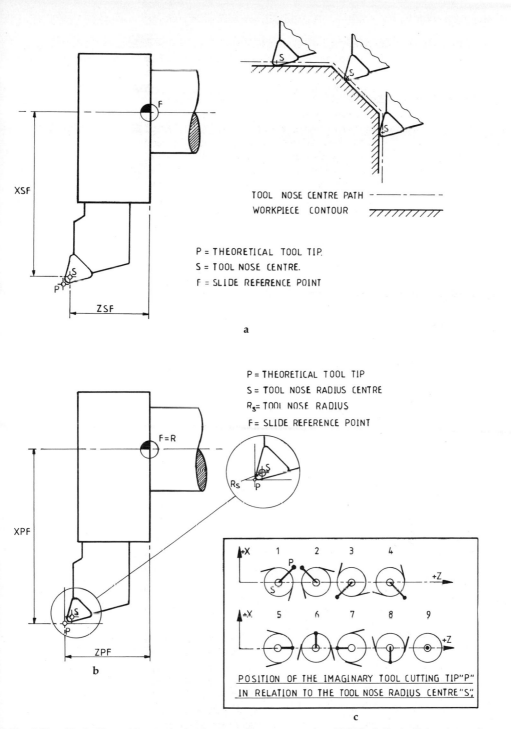

TOOL NOSE CENTRE PATH ------·--·--
WORKPIECE CONTOUR

P = THEORETICAL TOOL TIP.
S = TOOL NOSE CENTRE.
F = SLIDE REFERENCE POINT

a

P = THEORETICAL TOOL TIP
S = TOOL NOSE RADIUS CENTRE
R_s = TOOL NOSE RADIUS
F = SLIDE REFERENCE POINT

b

POSITION OF THE IMAGINARY TOOL CUTTING TIP "P"
IN RELATION TO THE TOOL NOSE RADIUS CENTRE "S"

c

Fig. 1.21. a Tool offset without using tool nose radius compensation (TNRC). **b** Tool offset using tool nose radius compensation (TNRC). **c** Position of the imaginary tool cutting tip "P" in relation to the tool nose radius centre "S". [Courtesy of Siemens.]

NB: With TNRC (i.e. G41 or G42) the difference, in addition to the tool nose radius, is traversed in BOTH axes, no axis command being required for traversing the tool offset.

Tool Offset Using Tool Nose Radius Compensation

It is possible to program a workpiece contour in conjunction with tool nose radius compensation (TNRC), as illustrated in Fig. 1.21b. The length of compensation entered into the controller is termed "the cutting point" and is designated by point "P". With this point known in terms of "X" and "Z" axes, the controller will then compute the tool path to be traversed and, as such, no contour error occurs. The engagement of TNRC takes effect at the end position of the block in which it is called (i.e. G41 or G42), meaning that in the following block the compensation is fully engaged.

In order to calculate and engage TNRC correctly, the controller requires a code indicating the position of the "imaginary" tool cutting tip "P" in relation to the tool nose centre "S" (see Fig. 1.21b), a magnified illustration for greater clarity is given.

Whenever the "XSF" and "ZSF" are selected as the tool dimensions instead of the "XPF" and "ZPF", namely the dimensions of the tool nose centre/slide reference points, one of the 9 codes depicted in Fig. 1.21c must be used for all the tools, with this particular controller. This is the convention for a turning centre with the turret situated above the workpiece; if the second (lower) turret is to be programmed the codes are applied in the same manner, with the only difference being that the X-axis direction is reversed.

Machining Centre, Selection and Cancellation of the Length Compensation

Only when either the G00 or G01 is active can the tool length compensation be selected. It is necessary to select the plane which is perpendicular to the length compensation direction, such as:

    N5 G00 G17 D.. Z.. LF

Only the tool's length compensation is written into the compensation memory using a "D-word"; this offset is always combined with the sign entered for its corresponding axis. To cancel length compensation, this is achieved with "D0", although the compensation is not removed unless its corresponding axis has been programmed. The following examples illustrate the effect of cancelling compensation with/without cutter radius compensation:

    length compensation without cutter radius compensation:
    N5 G90 G00 G17 D1 Z.. LF
    .

    .
    N50 D0 Z.. LF
    where: N5 = selection of length compensation (e.g. drill)
             N50 = cancellation of length compensation

    length compensation with cutter radius compensation:
    N5 G90 G00 G17 G41 D2 X... Y... LF
    N10 Z... LF

.
.
N50 G40 X... LF
N51 D0 Z... LF
where: N5 = automatic selection of cutter radius compensation

N10 = with length compensation
N50 = cancellation of cutter radius compensation
N51 = cancellation of length compensation

Intersection Cutter Radius Compensation

The compensation of the cutter radius is effective in any chosen plane, G16–G19, with the length compensation of the cutter always perpendicular to the selected plane, as we have seen previously. The G-codes for radius compensation are determined as follows:

G40 – no intersection cutter radius compensation
G41 – direction of tool travel to the left-hand side of the workpiece
G42 – direction of tool travel to the right-hand side of the workpiece

Whenever "mirror-imaging" is used, the path that might be travelled by the tool is depicted as follows, whilst taking the sign into consideration:

	Both axes mirrored or both axes without mirroring (Sign of cutter radius compensation value)		One axis mirrored	
	+	−	+	−
G41	Left	Right	Right	Left
G42	Right	Left	Left	Right

Selection/Cancellation of the Intersection Cutter Radius Compensation

Tool compensation preparatory functions can only be selected if either a G00 or G01 is active. It is possible to program G40, G41, G42 in a block which does not contain tool path moves, providing the rapid/feed function has been previously programmed in at least one axis. The following example highlights this point:

N10 G01 G17 G41 D07 X... Y... LF
N15 Z... LF
   where: N10 = at the end of this block the compensated path is reached in the selected plane – only the radius compensation value is incorporated
          N15 = the tool length compensation has now been incorporated into the program as well.

Yet another technique for introducing compensation into the program is depicted below:

N10 G17 LF
N15 G41 D07 LF
N20 G01 X... Y... LF
N25 Z... LF
    where: N10 = selection of the X–Y plane
           N15 = selection of compensation
           N20 = only the radius compensation has been selected at the end of
                 this block
           N25 = the tool length compensation has now also been incorporated.

Either cutter radius compensations, G41/G42, can be cancelled using G40, providing they are in linear blocks, G00/G01. So that the correct compensations are retracted, at least one plane must be programmed, with the length compensation being cancelled using "DO". This is, of course, assuming that its compensated axis was programmed. In the example given below, cancellation of cutter compensation is shown:

N30 G40 X... LF
N35 D0 Z... LF
    where: N30 = cancellation of tool compensation – only the radius compensa-
                 tion value is retracted
           N35 = the length compensation value = 0, is retracted.

Whilst not strictly a tool cancellation in the traditional sense, it is possible to swap one type of compensation for another, as the following change from G41 to G42 shows:

N10 G01 G17 G41 D12 X... Y... LF
N15 Z... LF
N20 G42 X... Y... LF
N25 Z... LF
    where: N10 = incorporation of the radius compensation to the left-hand side
                 of workpiece
           N15 = incorporation of the length compensation
           N20 = radius compensation changed to the right-hand side of
                 workpiece, for example, when changing the direction of
                 motion to the workpiece – traverse milling
           N25 = no change in tool compensation

Finally, in the examples of offsetting tooling, it is possible to change the tool offset number, without the need to re-enter the G-function (as it is modal), as the following examples indicate:

N10 G01 G17 G41 D12 X... Y... LF
N15 Z... LF
N20 D10 Z... LF
N25 X... Y... LF
    where: N20 = change in length compensation
           N25 = change in cutter radius compensation

Whenever the cutter radius compensation has been selected, it is generally not permissible to program either G58, G59, or G33 in most controllers. The remedy for this is to program these functions before selecting the appropriate cutter radius compensation, or alternatively, cancel the cutter radius compensation – select G58 .. G33 – then select the cutter radius compensation again. If the cutter radius compensation has been selected, including the G40 block, the effective zero offset value must not be changed.

In the following milling examples we can gain an appreciation of the advantages of utilising tool compensations, as only the part's dimensional features need be considered, once the appropriate compensation/s has been engaged.

Example 1. Milling a profile utilising cutter radius compensation (see Fig. 1.22a):

    N1 G01 G41 D1 G90 G17 X30 Y90 F500 S56 M03 LF
    N2 G91 X30 Y30 LF
    N3 G02 X30 Y-30 I0 J-30 LF
    N4 G01 X30 LF
    N5 G02 X30 Y30 I30 J0 LF
    N6 G01 X-15 Y-30 LF
    N7 X15 Y-30 LF
    N8 X-30 LF
    N9 X-30 Y-30 LF
    N10 X-45 Y30 LF
    N11 X-15 Y30 LF
    N12 G40 G90 X0 Y90 LF
    N13...

NB: As we can see (Fig. 1.22a), the milling cutter used a radius of 14 mm, with the cutter radius being entered under the tool offset number D1.

Example 2. Milling a circle utilising cutter radius compensation with scrolling – in/out – to avoid dwell marks (i.e. "witness") on the component (see Fig. 1.22b):

    N1 G90 G00 G17 G41 D1 X80 Y30 LF
    N2 G03 X130 Y80 I0 J50 LF
    N3 G91 G02 X0 Y0 I50 J0 LF
    N4 G90 G03 X80 Y130 I-50 J0 LF
    N5 G00 G40 X70 Y80 LF
    N6...

There is a whole host of virtually infinitely variable programming techniques that could be adopted in the machining of components. Just look how many there are and how they can vary. Even with individual part programs, there are many methods of programming motions – assuming the cutters take the same paths around the workpiece. If one decided to write a program using different tool motions for the same part, this would yet again produce considerable diversity in programs. There is no unique method for part programming a workpiece, only reasonable solutions, although it should be said that some programs can considerably reduce non-productive motions whilst machining the part, which can save seconds, or indeed minutes from the cycle time, depending upon how long, or complex the part. Not only can a more productive throughput result from optimum programming, but the length of blocks used in the memory might also be reduced, which is a further saving. A good part programmer can save a company considerable expense in reducing redundant work– non-productive cycle time to a minimum.

In the final section concerning the fundamentals of part programming we will consider further the effects of cutter and tool nose radius compensations, as their engagement needs to be fully understood and appreciated.

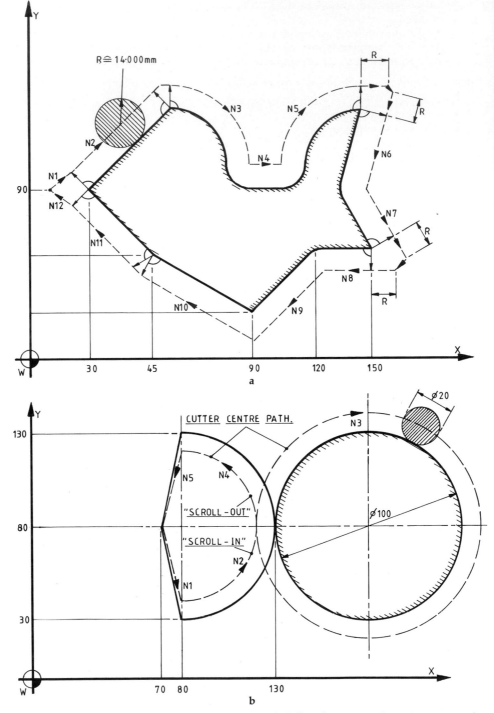

Fig. 1.22. a Milling with cutter radius compensation. **b** Full circle programming using cutter radius compensation. [Courtesy of Siemens.]

STRAIGHT LINE — angle "a" >180°

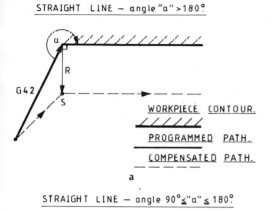

a

WORKPIECE CONTOUR.
///////

PROGRAMMED PATH.

COMPENSATED PATH.

CIRCULAR ARC — angle "a" >180°

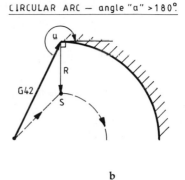

b

STRAIGHT LINE — angle 90°≤"a"≤180°

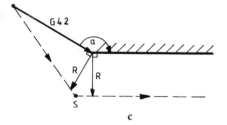

c

CIRCULAR ARC — angle 90°≤"a"≤180°

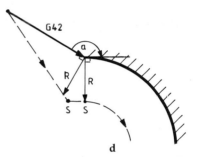

d

STRAIGHT LINE — angle "a" < 90°

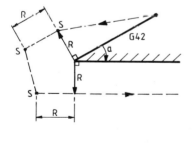

e

CIRCULAR ARC — angle "a"< 90°

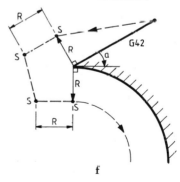

f

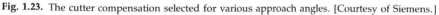

Fig. 1.23. The cutter compensation selected for various approach angles. [Courtesy of Siemens.]

1.4.13 Further Information on Cutter Radius Compensation (CRC)/Tool Nose Radius Compensation (TNRC)

The Selection of CRC/TNRC

We have seen how the compensation mode is selected for either CRC or TNRC previously (section 1.4.12) in the fixed plane, using either preparatory functions G41/G42, together with an offset number "D". The cutter compensation will either be to the left-hand side of the workpiece contour, in the traversing direction using a G41, or to the right-hand side of the component's contour with a G42. When selecting either CRC/TNRC, it is necessary for the controller to "look-ahead" – read two or three blocks ahead – in order to calculate the point of intersection. In the pictorial representations that follow, all the stop positions for single blocks are denoted by an "S", although we appreciate that in reality no such condition actually occurs in practice as we cut the part. In the following selection of typical cutter compensation engagements, a block start vector, denoted by the character "R", is created perpendicular to the programmed path, see Fig. 1.23a–f.

Obviously for the successful engagement of cutter compensation – which is always a problem for people new to CNC programming – certain rules must be adhered to:

the selection of the compensation mode can only be achieved in a block programmed with an "active" (i.e. modal) G00, G01, or alternatively either G02 or G03

the tool number D0 is assigned to the 0 compensation value and as such, no compensation is selected

CRC/TNRC in the Part Program

In the following schematic and tabulated examples (Fig. 1.24a–c), when using the selected CRC/TNRC, we have already mentioned how the controller, of necessity, must read two further blocks in advance during processing of the current block in order to calculate the intersection point for the compensated paths. These diagrams (Fig. 1.24a–c) illustrate just how such compensation is achieved for the various transitions:

straight line–straight line
straight line–circular arc
circular arc–straight line
circular arc–circular arc

NB: These geometric transitions of the various tool path vectors have been grouped according to the workpiece's included angle, denoted by the letter "a".

It is obvious from such pictorial representations that in order to generate successfully from obtuse to acute corner geometries (Fig. 1.24), the cutter's vectored path around these workpiece intersection points, is, of necessity, quite a complex motion.

NB: In Fig. 1.24 for all cases and for the sake of simplicity, the cutter compensation shown activated is G42, which as we now know, having the cutter to the right-hand side of the workpiece is in the traversing direction. It should also by now be appreciated that if G41 transitions had been applied, the same vectored paths would have resulted, except that the workpiece would lie on the left-hand side of the cutter's path.

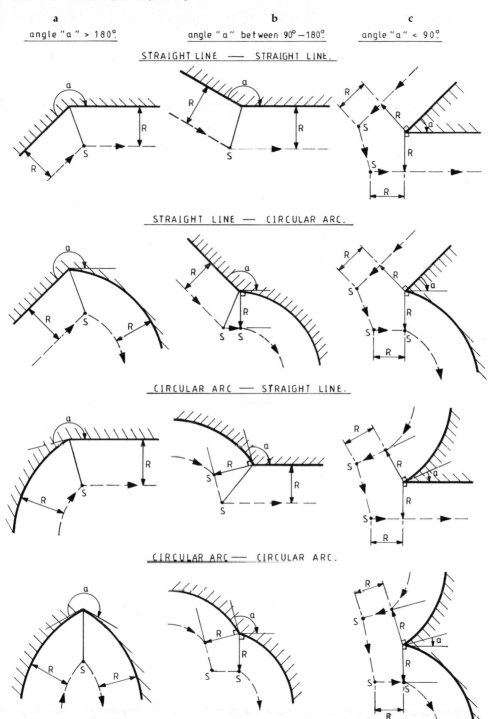

Fig. 1.24. Tool compensation for various workpiece geometric transitions. [Courtesy of Siemens.]

Cancellation of CRC/TNRC

Just as important as the successful engagement of cutter compensation is the cancellation of cutter compensation. In fact, it is even more critical to observe the rules of cutter compensation cancellation, than those for its engagement. This is because if we engage a cutter transition compensation incorrectly, the worst thing that can happen is that we will cause a vectored cutter path which might machine too much off the workpiece and lead to scrapping the part, or at best leave extra stock material behind – after the passage of the cutter – so that further machining would be necessary at some later stage. However, if cutter compensation cancellation is retracted at the wrong point in the part program, it can at worst cause a rapid sideways radial motion as it is cancelled. There is damage to both the workpiece and cutter, with the severity being dependent upon the magnitude of the cutter's radius. Therefore, extreme care must be used in determining the earliest point within the program at which cutter compensation cancellation can be successfully achieved.

It is apparent to the reader by now that the compensation mode is cancelled using the preparatory function G40. In the following selected examples (Fig. 1.25a–c), cancellation of cutter compensation is depicted for both straight line and circular arc tool path motions for varying workpiece included angles. As previously mentioned for the engagement rules of cutter compensation, the cancellation can only be achieved in a program block which is "active" with either G00 or G01 linear motions, or G02/G03 circular motions. Furthermore, the tool number D0 corresponds with a value 0, which allows it to be used to cancel the tool compensation also.

Changing the Direction of Compensation

There are many occasions when it is necessary to change the direction of cutter radius compensation within the part program, in order to machine a certain feature. A typical example of this is shown in Fig. 1.26a, where a top face has been milled and the compensation is changed to machine the chamfer. This diagram shows the perpendicular vector with a length "R" which is created in the appropriate direction of compensation at the end position of the old G-function block, in this case G42, and at the starting point for the new block with its respective G-function G41.

Changing the Offset Number

If we change the offset number at any point within the program, then the following logic applies (Fig. 1.26b):

there is no block start intersection calculated from the old compensation

a perpendicular vector with length "R1" is created at the end position of the block using the old offset number

the block end intersection is calculated with the new compensation value

Changing the Compensation Values

As the reader can appreciate, this modification to the radius compensation value (Fig. 1.26c) is similar in its function to that of changing the offset number described above (Fig. 1.26b). The compensation values may be changed at the:

a

angle "a" > 180°.

STRAIGHT LINE.

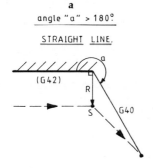

CIRCULAR ARC.

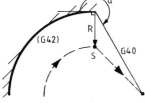

b

angle "a" between 90° to 180°.

STRAIGHT LINE.

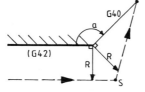

CIRCULAR ARC.

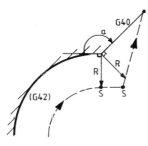

c

angle "a" < 90°.

STRAIGHT LINE.

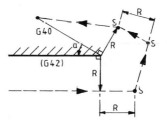

CIRCULAR ARC.

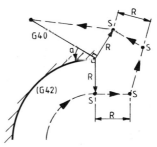

Fig. 1.25. Cancellation of tool compensation using the preparation function G40. [Courtesy of Siemens.]

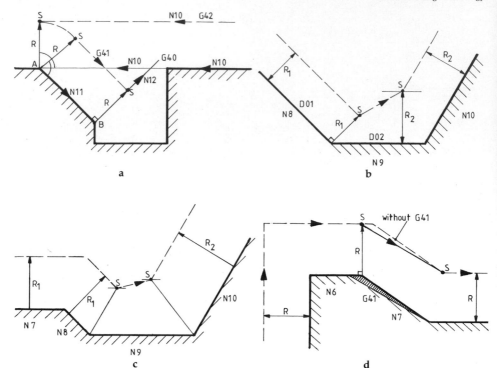

Fig. 1.26. Adjustment of cutter path. **a** Changing the direction of compensation. **b** Changing the offset number. **c** Changing the compensation values. **d** Repetition of the selected G-function (G41, G42) with the same offset number. [Courtesy of Siemens.]

operator's panel
using an external tool offset, or in the part program
tape reader – using an NC tape

The new compensation value takes effect in the next block of the part program.

Repetition of the Selected G-Function, with the same Offset Number

Assuming that either a G41, or G42 preparatory has already been programmed and is repeated, a vector with length "R" and perpendicular to the programmed path is created in the preceding block at the block end position (Fig. 1.26d).

The following example illustrates the block start intersection being calculated for the following block:

N4 G91 D10 G41 X...Y...LF
N5 Y...LF
N6 X...LF
N7 G41 X...Y...LF
N8 X...LF

.

NB: This extra G41 in block N7 being repeated from block N4, causes an error, thus, extra stock is removed and part is scrapped (Fig. 1.26d).

The Effect of Using M00, M01, M02 and M30 with either CRC/TNRC Selected

M00 and M01: the CNC stops, when this preparatory function is programmed, at the position "S" (shown in Fig. 1.26), for a single block.

M02 and M30: when these preparatory functions are active, the compensation is also retracted if it is cancelled in the last block with a G40, assuming that at least one axis address has been programmed. The following example shows the function M30 within the program:

```
.
N150 X...Y...LF
N200 G40 X...M30 LF
.
```

Obviously the compensation is not retracted if a cutter path has not been programmed.

CRC/TNRC with a Combination of Various Block Types and in Conjunction with Contour Errors

If one is programming in a contouring mode, special attention must be paid to the blocks without tool movement in order to prevent contour errors, as the following examples illustrate:

when tool path addresses are programmed, but there is no movement, since the distance is 0 – as the example below shows:

```
.
N...G91 X0 LF
.
```

auxiliary functions, such as a dwell, axis address outside the compensation plane, or a zero offset programmed in the compensation plane instead of path addresses, shown below:

```
.
N...M05 LF
N...S21 LF
N...G04 X100 LF
.
```

When one "auxiliary function block" is programmed between the tool paths in the compensation plane, no error occurs on the part:

```
N5 G91 X100 LF
N6 M08 LF
N7 Y-100 LF
.
```

However, when two "auxiliary function blocks" are programmed between the tool paths in the compensation plane, a corner error occurs. The part has an inappropriate tool motion, causing a chamfering to the corner:

N5 G91 X100 LF
N6 M08 LF
N7 M09 LF
N8 Y-100 LF
N9 X100 LF

NB: There are many examples that can be caused by incorrect usage of M-functions, leading to part scrappage by unanticipated and undesirable tool motions, so extreme care in when, where and how they are used, is urged.

Special Case CRC/TNRC Problems

In this final discussion about cutter compensation programming techniques, a range of "special case" problems for both CRC/TNRC will be highlighted. The first example chosen (Fig. 1.27a) illustrates how the controller logic always uses the next block to calculate the point of intersection of the compensated paths. Assuming that no axes in the compensation plane are programmed in the next block, the controller will automatically skip this block and use the following one. When this occurs there is a likelihood that a contour error will take place, if the intermediate block is less than the compensated value. Machining is not interrupted, although an alarm signal is indicated.

Fig. 1.27b shows the expected outcome of the toolpath motions when an intermediate block is too small for the selected compensation.

The illustration depicted in Fig. 1.27c shows the problem when the direction of cutter compensation of either CRC/TNRC is retained and the traversing direction is reversed. Note that the return path, shown by line N2, must exceed twice the cutter radius/tool nose radius, or the tool will proceed to move in the wrong direction.

The following diagrams (Figs. 1.27d–f) apply to external contours with circle transitions having obtuse angles:

Fig. 1.27d: in order to prevent a conditional stop in the contouring mode owing to intermediate blocks which are too small, the tool paths "AB" and "BC" can be omitted within the CNC

Fig. 1.27e: depending upon the tolerance "d" which is defined on start up – maximum being 32 000 μm, the path will be as follows:
if X1 and Y1 are less than "d", there will be a direct traverse from "A" to "C"

Fig. 1.27f: if X1, Y1, X2 and Y2 are less than "d", there will be no compensating movement and machining continues with a new radius at point "A", producing a machining error

The block numbers are interchanged whenever CRC/TNRC generates intermediate blocks – including those upon selection and cancellation – if an axis movement outside the compensation plane is programmed between these blocks.

In Fig. 1.27g, an indication of such programming problems can be seen, as the following logic shows:

N5 G00 Z100 LF
N10 X...Y10 LF
N15 G31 D01 X20 Y20 LF
N20 G03 X0 Y40 I-20 J0 LF

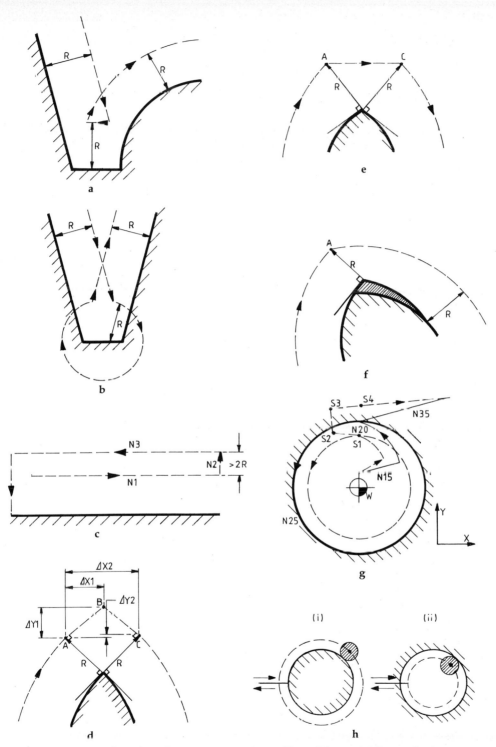

Fig. 1.27. "Special case" cutter compensation problems. [Courtesy of Siemens.]

N25 X0 Y40 I0 J-40 LF
N30 G01 Z0 LF
N35 G40 X80 Y60 LF

NB: The points S1, S2, S3 and S4 belong logically to block N25, with the machining sequence, visible in the single block, as follows:

. . . , N20, N25 (S1), N30 (tool withdrawn from the workpiece), N25 (S2), N25 (S3), N25 (S4), N35 . . . This sequence is also valid if N25 is a linear block.

Fig. 1.27h illustrates the effect of using negative compensation values. This means that a compensated path corresponding to a G42, with a positive compensation value, is implemented with a G41 – i.e. an analog internal contour is followed instead of the programmed external contour, or vice versa.

In Fig. 1.27h(i), the cutter centre path shown has a positive compensation value which has been entered, whereas in Fig. 1.27h(ii), a negative compensation value is illustrated in conjunction with the same machining program. If the program is generated as shown in Fig. 1.27h(ii), with a positive compensation value, it will have a negative compensation effect and produce machining as described in Fig. 1.27(i). Therefore, it is possible to implement two machining conditions using the same program and they are distinguished by entering either a positive or negative compensation value.

In the final section concerning CNC programming fundamentals we will consider the advantages to be gained from utilising programming aids termed "canned cycles".

1.4.14 Part Programming Using "Canned Cycles"

Canned cycles consist of a predetermined series of machining operations that direct the movements of all axes and the spindle. Canned cycles are intended to permit operations such as drilling, tapping, boring, pocket clearances and so on, without requiring repetitive programming of all data in each block of information. Many controllers have over ten canned cycles which can be called upon by the programmer. This allows the user to select the actions desired for these canned cycles. Each cycle is defined as a multi-step sequence of operations, with each step representing a departure mode, switching mode, or a mode of operation. For example, with a typical controller up to 28 actions are selectable allowing the programmer to customise a canned cycle, with up to 24 steps being used to define a canned cycle. In the following list, there is a typical range of actions available which can be user defined on a vertical machining centre:

    00 – end of cycle
    01 – rapid to R-plane
    02 – start spindle
    03 – stop spindle
    04 – orientate spindle
    05 – reverse spindle direction
    06 – off-centre position
    07 – remove off-centre position
    08 – feed to depth
    09 – incremental feed to depth
    10 to 13 – dwell cycles
    14 – return to R-plane at traverse rate

15 – return to R-plane at feedrate
16 – return to initial position at traverse rate
17 – return to initial position at feedrate
18 – programmed spindle direction
19 – turn on operator's feedhold
20 – distance to zero
21 to 28 – disable/enable and dwell cycles

NB: The programming manuals would offer a detailed description of each programmable action and how to engage them within the canned cycle.

In the following typical canned cycles for machining centres, the reader can gain an appreciation of how they operate. The first canned cycle we will consider is the standard drilling cycle, depicted in Fig. 1.28a:

rapid traversing to an X/Y position within the program
rapid traverse to a preselected R-plane, above which rapid motions can be safely made
feeding to the programmed Z-depth at a preselected feedrate
return to R-plane at feedrate, this being the end of the canned cycle
rapid traverse to the next X/Y position

In the second example of canned cycle programming this is used in conjunction with the drilling cycle (Fig. 1.28a) and is a spotfacing operation with a dwell (Fig. 1.28b):

rapid traverse to drilled hole at the X/Y position
rapid traverse to the pre-selected R-plane
feeding to programmable Z-depth
dwelling to clean up counterbore
return to R-plane at feedrate, end of canned cycle
rapid traverse to next X/Y hole position

After drilling the holes in the component using a canned cycle, it is often desirable to incorporate a tapping cycle (Fig. 1.29a) as a nested subroutine within the main program. A typical tapping canned cycle is shown in Fig. 1.29a:

rapid traversing to the X/Y position
rapid traverse to the preselected R-plane
feeding to required Z-depth
reversal of spindle rotation and return to R-plane
cancel spindle reversal and stop spindle
rapid to following X/Y hole position, as necessary

The final canned cycle, depicted in Fig. 1.29b, is a boring cycle having a dwell with feedrate return:

rapid traverse to hole position in X/Y plane
rapid traverse to R-plane
feeding to specified Z-depth
timed dwell
return to R-plane at feedrate
traverse at rapid to next hole position in X/Y plane as necessary

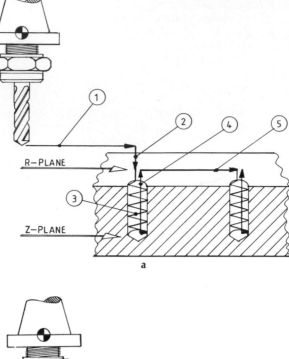

a

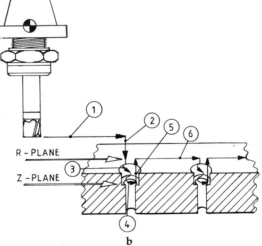

b

Fig. 1.28. Using "canned cycles" for drilling and spotfacing operations. **a** A standard drilling cycle (i.e. G81). 1, traverse to position (X/Y); 2, traverse to R-plane; 3, feed to programmed Z dimension (i.e. in-zone position); 4, return at traverse to R-plane (end of cycle); 5, traverse to next position (X/Y). **b** Drilling cycle with dwell – "spotfacing" (i.e. G82). 1, traverse to position (X/Y); 2, rapid to R-plane; 3, feed to programmed Z dimension (i.e. in-zone position); 4, dwell time; 5, return at traverse to R-plane (end of cycle); 6, traverse to next position (X/Y). [Courtesy of GE Fanuc.]

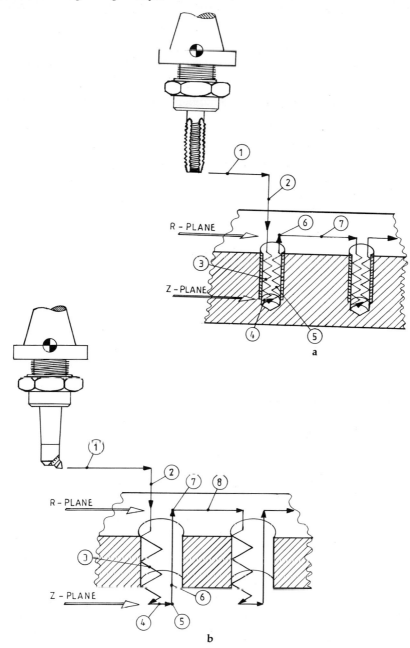

Fig. 1.29. Typical "canned cycles" for tapping and boring operations. **a** Tapping cycle (i.e. G84). 1, traverse to position (X/Y); 2, traverse to R-plane (i.e. in-zone position); 3, feed to Z dimension (i.e. distance zero position); 4, reverse spindle for return to R-plane; 6, remove spindle reversal and enable feedhold (end of cycle); 7, traverse to next position (X/Y). **b** Boring cycle with spindle stop, traverse return (i.e. G86). 1, traverse to position (X/Y); 2, traverse to R-plane; 3, feedrate to Z dimension (in-zone position); 4, dwell; 5, spindle stop; 6, return at traverse to R-plane (in-zone position), 7, spindle start (end of cycle); 8, traverse to next position (X/Y). [Courtesy of GE Fanuc.]

A User-Macro

Such individual machining operations as those listed above can be incorporated into larger canned cycles, often termed a "user-macro". Typical user-macros are shown in Fig. 1.30a,b for a linear and circular drilling pattern, respectively. By definition, a user-macro is used to repeat a series of actions at several programmed positions. The repeated actions are defined in the programming logic and once called, the execution of the series of blocks containing axis moves causes the macro to re-execute these repetitive operations until either cancelled, or another canned cycle is programmed.

In Fig. 1.30a, the linear drilling pattern consists of a row, column, or rectangular grid of hole centres which can lie at any angle from $-360°$ to $360°$ relative to the horizontal axis – obviously on a vertical machining centre. In generating a drilling pattern, the machine tool traverses to all the defined hole centres in order, including any "don't drill" points. The currently active canned cycle is then executed at each hole centre unless it is defined by a "don't drill" point.

In Fig. 1.30a, the general form of linear pattern programming is to activate the desired canned cycle, then call for a linear pattern as depicted in the linear pattern of holes, where:

    XY = start point of the pattern
    IJ   = signed increments from the start point to centre
    P2 = signed incremental distance between holes
    P3 = 3 rows
    P8 = last row of holes from first
    P9 = row spacing – last hole

A circular drilling pattern (Fig. 1.30b) consists of a set of equally spaced holes positioned around the circumference of a circle of given centre and radius. In generating the hole pattern the machine traverses to all the defined points in order, including those defined as "don't drill" points. The currently active canned cycle is then executed at these points unless defined as "don't drill" points. The general form for programming a circular hole pattern is to activate the desired canned cycle, then call the circular pattern (Fig. 1.30b), as follows:

    where G81 = standard drilling cycle
           XY   = start point of the pattern
           IJ   = signed increments from the start point to centre
           P2   = signed incremental angular distance between holes
           P4   = angle of last hole in pattern
           P9   = finishing position indicator

NB: There is a whole host of different linear and circular drilling patterns that can be programmed and those just mentioned were only included to illustrate how the program logic for such specific holes is structured. Furthermore, this is not intended by any means to be an exhaustive account of the permutations of holes that it is possible to program within the part program, or indeed of any of the dimensional features that can be logically described using either "word address" or "parametric" programming. The intention with section 1.4 was simply to give the reader an appreciation of the programming aids and logical structures that may be called upon when writing part programs.

The final sections of this chapter will examine how a range of specialised CNC applications can be incorporated into such machine tools, which "open-up" the programming and manufacturing opportunities, further increasing the diversity of

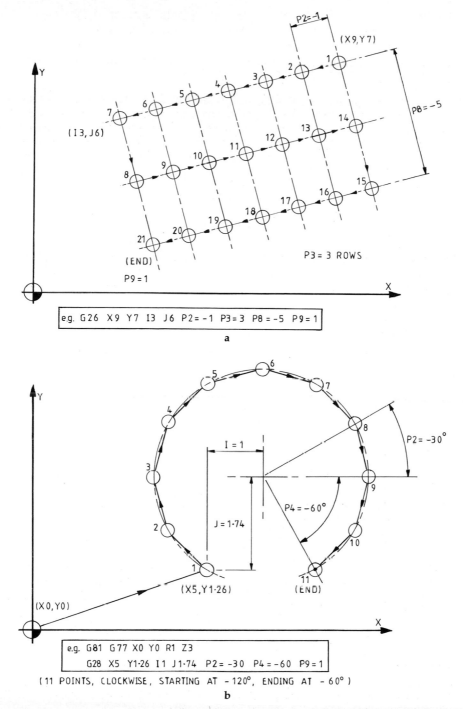

Fig. 1.30. Typical "canned cycles" for drilling. **a** Linear drilling pattern (G26). **b** Circular drilling pattern (G28). [Courtesy of GE Fanuc.]

such CNC machines. The importance of such applications will become increasingly popular as companies come under commercial pressures in the years to come.

1.5 High-speed Milling Fundamentals

With the new developments in carbide, ceramic, polycrystalline diamond and cubic boron nitride tooling, cutting speed potential has dramatically increased on both turning and machining centres. Such tooling developments have led to the construction of high-speed spindles, improved bearing design and lubrication systems, advances in spindle cooling equipment, together with more rigid machine tool structures, and as such, allowing the exploitation of higher cutting speeds. Improvements in productivity, together with the elimination of chatter and longer tool life, have been the primary objectives of these developments. This has resulted in the current "state-of-the-art" machine tools, where the contouring speed is the limitation for profiling accuracy requirements, or the CNC's processing speed, rather than the physical metal removal rates.

In any high-speed machining operation, the principal factors affecting tool path accuracy, and hence the part geometry, are generally considered to be:

basic construction of machine tool, in particular its rigidity and accuracy

cutter design and stiffness

servo-lag

control processing speed

In the following sub-sections, we will consider each of these factors in turn along with their influence in a high-speed milling situation particularly.

1.5.1 Machine Tool Rigidity and Accuracy

In this section we will consider only two pertinent points: first, that high-speed milling is a dynamic process that can cause much higher stresses on the machine tool than would occur in traditional machining operations. This means that the machine, of necessity, must be much more rigid so that it can absorb these higher stresses without causing unacceptable deflection of its basic design, whilst simultaneously increasing the "dampening effect". Secondly, precise contour milling operations depend upon the basic machine tool accuracy. In the following sections we will discuss the factors that are unique to the generation of contoured surfaces at high speeds. The reader also needs to appreciate that the straightness and alignment of the axes, errors in positioning, plus the repeatability errors will be superimposed onto the finished part in addition to any errors that may be created by the dynamics of high-speed milling.

1.5.2 Cutter Design and Stiffness

Milling cutters can be designed that will run up to and over speeds of 40 000 r.p.m., but this is beyond the scope of this discussion. Their design must ensure that large chip gullets occur and that the cutter can be dynamically balanced in the radial and axial planes – dual plane balancing – to reduce vibrations at high speed and damp such tendencies, with cutter stiffness rigidity as a high priority. Normally, when any high-speed contour milling is necessary, the use of relatively small diameter cutters becomes desirable to reach into intricate three-dimensional surface features and as such they can easily be deflected. If we change either the feedrates or the amount of stock to be removed, this affects the cutter by varying the forces, which in turn influences its deflection. Cutter deflection may adversely affect the machine tool's ability to reproduce the programmed contour faithfully but owing to high speeds deflections are minimised.

One must always keep in mind that any potential cutter deflection is a function of the material, geometry and its length-to-diameter ratio. For example, a cemented carbide cutter is of the order of three times stiffer than its equivalent high-speed steel cutter, having the same size and geometry. We have already seen in chapter 2 that cutter deflection – when manufactured from the same material – does not vary as a linear function of its length but as a cubic function of it; therefore a 50 mm long cutter will potentially deflect eight times more than a 25 mm cutter operating under the same load. This leads to an obvious recommendation to utilise the shortest acceptable cutter to machine the part features, made from cemented carbide with "dual plane balancing" and having good chip evacuation abilities.

1.5.3 Servo-lag Problems Affecting the Machined Contour

Today, most CNC machine tools use "proportional servo-systems", where the axis velocity is proportional to the difference between the actual position and the command position (Fig. 1.31a). This "error signal" is used by the system to determine any acceleration/deceleration necessary as well as the steady-state velocities. As one can appreciate from Fig. 1.31a, the distance between the actual and commanded positions is commonly termed "servo-lag". This is taken a stage further in Fig. 1.31b, where the illustration depicts how a "proportional servo-system" is used to mill a sloping line. In this example, DX and DY are the total programmed changes in position on the X and Y axes, respectively, to go from point "A" to point "B", whereas DX_L and DY_L are the amount of lag on each axis at point "C" along the tool path from "A" to "B". Furthermore, in such a system the lag on the X-axis must be proportional to a similar lag on the Y-axis, in order to accurately follow the slope of the line. This can be represented mathematically by the following relationship:

$$\frac{DX_L}{DY_L} = \frac{DX}{DY} = \text{slope of the line}$$

In Fig. 1.31c, we can gain an appreciation of just what happens when the servo-lag on both axes is not proportional. As the machine travels from point "A" to point "B", the lag on the X-axis is proportionally less than the lag on the Y-axis. This might be the result of the servo-gains between the X- and Y-axes not being properly synchronised. Incidentally, "gain", or servo-gain in this case, is a measure of the servo's responsiveness – with the higher the gain, the lower the lag. Normally, gain can be expressed in mm/min (i.e. velocity)/mill (i.e. distance in 0.001) of lag. Lag can be found using the following relationship:

$$\text{Lag "L" (mm)} = \frac{\text{Feedrate (mm/min)}}{\text{Gain "G" (mm/min/0.25)}}$$

For example, if a slide is travelling at 2500 mm/min and the servo has a gain of 2, the lag will be 1.25 mm as shown by the calculation:

$$L = \frac{F}{G} = \frac{2500}{2/0.001} = 1.25 \text{ mm}$$

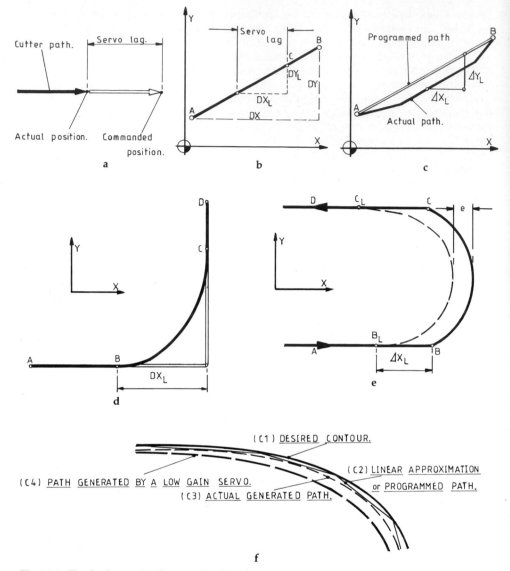

Fig. 1.31. The fundamentals of precision in high-speed milling operations. **a** A simple example of servo-lag. **b** How a proportional servo system mills a sloping line. **c** What happens if servo-lag on both axes is not proportional. **d** The effect of servo-lag and gain on corner milling. **e** The effect of servo-lag and gain on circular paths. **f** The impact of servo-lag when following a contour. [Courtesy of Boston Digital Corporation.]

The Effect of Servo-lag and Gain on Corner Milling

If two axes with correctly matched servo-lag can move in a straight line from point "A" to point "B", then to understand the effect of gain, let us consider what occurs when milling a right-angled corner at a constant feedrate without stopping (Fig.

1.31d). Whilst cutting the corner from "A" to "B" and then onward to "D", the servo develops a steady lag (DX_L), until sufficient command signals have been generated to reach point "B". It is at this position that the control begins to generate commands toward point "D", although the actual slideway has not yet reached point "B", owing to the servo-lag (DX_L). At this point the X-axis will begin to decelerate and, simultaneously, the Y-axis begins to accelerate, i.e. the velocity is proportional to the distance between the command signal and the actual position. It is not until point "C" is attained that the X-axis slide actually stops. Acceleration factors affect the slideway motions producing the result that the distance from "B" to "C" is always greater than DX_L. Furthermore, this is not a circular arc, but an exponential curve, with the amount of variance from the sharp right-angled corner being dependent on the magnitude of servo-lag, which itself depends upon the affect of feedrate and gain – according to the previous formula.

The Effect of Servo-lag and Gain Whilst Generating Circular Paths

For the reader to understand just what happens in milling complex contours, we will consider the case of two straight lines joined by a semi-circle (Fig. 1.31e). In this case, milling occurs at a constant feedrate from point "A" in a straight line until the command dimension reaches point "B". However, at this point, because of the effect of servo-lag, the slide will have only reached point "B_L". Therefore, as the control command is moving forward at a constant rate, it begins to generate commands toward point "C". This results in the slide beginning to move away from the desired path at point "B_L". The dotted line shows the actual path taken by the cutter and as we can see, from points "B_L" to "C_L" the deviation from the desired path is shown as "e".

In this example the magnitude of "e" is determined as a function of the feedrate and gain and the desired radius. When the radius error approaches the programmed radius, the resulting machined profile appears distorted and is hence impracticable. If one wanted to cut a 25 mm radius at a feedrate of 2500 mm/min with a machine tool gain being 25 mm/min/0.001, then the error generated would be approximately 0.125 mm; whereas if the gain is increased to 100 mm/min/0.001, the maximum error "e" will be reduced to approximately 0.008 mm.

A machined curve is an approximation on CNC machine tools, in that the profile is constructed from a series of short connected segments, or chords. The controlling factor on the length of such segments is the deviation between the centrepoint of any chord and a point at right angles on the programmed curve. The linear distance between these two points is usually termed the "maximum allowable chordal deviation" and is a function of the controller's executive software. Therefore the resultant machined curve is a combination of the chordal deviation and the servo-lag for a particular machine tool.

Illustrated in Fig. 1.31f, is the culmination of servo-lag when following a contour, with the curve "C1" being the desired contour, "C2" a linear approximation (the programmed path), "C3" the actual generated path resulting from servo-lag utilising a high gain servo and, finally, "C4" being the path generated by a low gain servo. Through servo-lag a smoothing of any contour occurs owing to the lagged cutter path; this causes severe contour problems with respect to part accuracy for simple arcs, as shown in Fig. 1.31e.

Clearly, servo-lag and gain promote a variety of effects on complex shapes depending upon their geometry and tolerance and these effects become still more complicated when considering three-dimensional contouring. In many circumstances, the cutting of three-dimensional profiles may necessitate utilising four or five axes of movement to produce the part. The servo-lag and gain on all axes must be considered when manufacturing complex and accurate parts. Regardless of part complexity, or indeed the number of axes utilised, there is one point that should be emphasised: potential errors created by servo-lag can exceed the errors in the basic positioning accuracy specifications for any machine tool.

1.5.4 CNC Processing Speed

Probably the main factor limiting contouring speed is the processing speed of the CNC, with each "stroke" generated for every axis which must be read, interpreted and activated. This is usually referred to as the "block processing time". The maximum allocated time for block processing of information is dependent on the length of the stroke and the feedrate. It is possible to calculate the maximum block processing time (T_b) as follows:

$$T_b = \frac{\text{Maximum stroke length}}{\text{feedrate}}$$

For example, if we require a chord length, i.e. stroke length, of 0.50 mm, in order to maintain contouring accuracy whilst milling at 3000 mm/min, or 50 mm/s, with the maximum block processing time it should be less than:

$$T_b = \frac{0.50}{3000/60} = \frac{0.50}{50} = 0.01 \, s, \text{ or } 10 \, ms$$

Many CNCs have block processing times within the range of 60–80 ms, as we can see in this case the program would suffer from "data starvation", whilst the controller caught up on its data processing. Such "starvation" would cause hesitation in the slides, slow down the cutting time and leave dwell marks on the workpiece. Since this is unacceptable, a lower feedrate must be programmed and a longer cycle time will result. In this example, if the block processing time of the controller is 60 ms, the cut would take six times longer to generate the profile than a controller having a processing time of 10 ms.

In order to understand more fully the problems mentioned above, we will consider two widely differing applications, in the first instance the milling of a hob to manufacture a die used in producing intricate metal buttons. Such a hob will more than likely have fine detailed work on it, with radii as small as 0.25 mm requiring a tool tip radius of 0.025 mm. In order to machine features with such a small cutter, spindle speeds might reach 40000 r.p.m. utilising feed per revolution of 0.008 mm giving a feedrate of 320 mm/min. Many people would not consider this as high-speed milling, but let us look more closely at this particular problem. If the controller has a servo gain of 4, with a feedrate of 320 mm/min, this means that the servo-lag would be 0.75 mm/min, which is consistent with producing radii of 0.25 mm/min. However, if the gain was one, this would cause a servo-lag of 0.320 mm/min and in this case it obviously could not machine the part. In such circumstances it would be necessary to reduce the feedrate to 75 mm/min to generate the contour and this means the cutting time increases by a factor of four.

Let us now consider the impact of block processing time under these conditions. To cut a radius as small as 0.25 mm, we would need to produce linear stroke lengths of 0.075 mm to reproduce acceptable detail. This requires a block processing of 15 ms. If the controller has a block processing time of 60 ms, then the feedrate must be limited to 75 mm/run which increases milling time by a factor of four.

The second example to be considered is the casting pattern for a large ECM electrode for a turbine fan, with the material being aluminium having very gentle three-dimensional curves. In this case, the spindle has a 250 000 mm/min capability and with adequate power to cut at a feed of 0.25 mm/rev. This would indicate a feedrate of 62 500 mm/min (i.e. 250 000 × 0.25 = 62 500 mm/min) would be possible. For accuracy, a chordal deviation (C_d) of 0.005 mm would indicate a stroke length of 0.75 mm if the minimum radius of curvature was 25 mm.

Assuming that a servo gain of one was available, then we would get errors as large as 0.125 mm; this would not produce an acceptable part; also at 62 500 mm/min, a block processing time (T_b) of 60 ms would require stroke lengths of 2.5 mm instead of the 0.75 mm we needed for the required accuracy. Therefore, in order to eliminate the effects of low gain or slow processing time, it is necessary to depress the feedrate, resulting in cutting time increased up to 400%.

In considering these two examples metaphorically, one method is like racing a go-cart on a small tight track, whilst the other is similar to a highly tuned sports car on a longer and smoother track. The go-cart may only reach 30 km/h, whereas the sports car may hit 200 km/h. The corner forces and the reaction times are similar, even though the speeds are vastly different. Looked at yet another way, we can say that the frequency of response of drive and car, i.e. servo gain and processing time, are similar in both examples even though the speeds (feedrates) are radically different.

In the day-to-day production environment, the duplication of specific and precise contours is the end result of a combination of inter-related factors. As the number of axes required to produce the part increases, the difficulty of obtaining the desired shape will increase proportionally. So, machine tools that produce excellent general purpose work may not be either accurate enough or efficient enough when machining contours. Therefore the machine tool described and partially illustrated in Figs. 1.31 and 1.34e, respectively, has been specifically designed so that that block processing time is as low as 10 ms, instead of the usual 60–80 ms. With this servo system, gains up to 4 instead of the more common 1 or 2 are provided.

This completes the review of the implications of high-speed milling operations and many of these problems of both the effects of servo-lag and gain are true in turning operations, with the exception being that workholding becomes the real cause for concern, as the following section shows.

1.6 High-speed Turning Operations

As was suggested above, servo-lag and gain are also crucial in any high-speed turning operations, as response time and data processing speed become paramount to any instantaneous vectoring of the tool around the workpiece, together with an advanced "look-ahead" capability.

However, turning operations at such ultra-high speeds differ fundamentally from high-speed milling, in that the workpiece revolves and this in itself is the major factor which must be addressed. When headstocks are rotated at or above 12 000 r.p.m. there is a limitation when using conventional contact bearing techniques for spindle rotations. It becomes necessary to incorporate air-bearing spindles, preferably with direct-drives that have the added benefit of removing from the system the transmission problems of conventional drives which can influence part geometrical features. Any workpieces having out-of-balance non-symmetrical features should be avoided, as their dynamic balance at speed will not only become a problem in terms of destabilising the cutting process, but can cause problems with safety when work-holding the part – possibly throwing the component from the chuck, between centres, expanding mandrel, etc.

The workholding techniques utilised can influence yet more problems, such as "bursting pressures" associated with centrifugal forces affecting the chucks, etc., so as the internal forces build up with rotational speed, if high-strength materials are not used then such devices can literally explode. When long, thin workpieces are revolved at high speed there is a tendency to "whipping" which may cause either damage to the machine tool or affect operator safety. There are many more such problems occurring owing to high speeds which do not readily arise at conventional rotational speeds, but a full description of such problems was not the intention here, only an appreciation of the influence of ultra-high speeds during turning operations.

1.7 "Reverse Engineering" – an Overview of Digitising on Machining Centres

Touch-trigger probes have been in general use on machining centres for well over ten years and during this period the process-related metrological applications of workpiece setup and tool wear compensation have occurred. A touch-trigger works by passing signals to the controller to record the points when the stylus makes contact with the workpiece and in so doing stops the machine's slide motions. Furthermore, a number of these points can be used for computing workpiece positions, diameters and angles. Any controller has to be able to accept an external signal as an interrupt to its motion and act accordingly with computation being possible whether it is instigated by the programmer, within the program, or by the controller's executive functions. Such touch-trigger probes are offered by the vast majority of machine tool builders and as such they form a basis for the digitising of the workpiece in the following discussion.

1.7.1 The Principles of Digitising

With any digitising technique it is necessary to position a model of the part to be digitised within a predefined envelope, or area, which is then scanned at discrete intervals. The positional data obtained from scanning the model can be used to construct a replica. The reproduction quality depends upon several factors, most of which are sourced through the digitising method chosen, whilst other influences include the software capability when building the model's replica.

It should be quite feasible to build a three-dimensional model using a mainframe computer, although a more limited two-dimensional model could be built with a desk-top computer. It is necessary for the system designer to choose either two or three dimensions at an early stage, since the cost of both computer and its respective software has a significant bearing on its cost-performance factors. Briefly then, to describe the differences between the two systems, it is necessary for the reader to visualise a wire-frame model (see Fig. 1.33c). In a two-dimensional system the computer can only be aware of the intersections on the wire frame, knowing nothing about the space between the points, whereas a three-dimensional system uses these points to construct surfaces, and as such knows a great deal more about the model's geometry. At present, the market values of cost ratios between these two systems is about 10:1, in terms of software costs alone – in favour of the two-dimensional system, of course. This two-dimensional system forms the basis for discussion here, with occasional reference to three-dimensional techniques.

Fig. 1.32a(i) shows a touch-trigger probe fitted with a stylus of a known diameter. It is shown in contact with a model being digitised at a given position in the X–Y plane and finding the Z-plane for that position. In order to cut an exact replica of the digitised model, a tool of identical proportions to the stylus (Fig. 1.32a(ii)) can be positioned by the machine tool to the same coordinates, reproducing the point of contact. Once the stylus is in contact with the workpiece surface, a series of points can be digitised and the coordinates of the surface contacted may be linked to form a cutter path (Fig. 1.32b) using these consecutive points.

The probe's path – hence the cutter path – is formed by defining a grid over which the probe must move (Fig. 1.32c) and by fixing this grid, two of the three coordinates are automatically known leaving the third to be captured by the probe. The grid must be defined in such a manner that the model (or relevant areas to be digitised) is covered. Therefore the probe moves to each point in succession and captures the data by its motion along a grid line automatically fixing one axis. The second axis coordinate is found by moving along that line in discrete steps, whereas the third axis is found by digitising at that point.

Using this basic principle, described above, it is manipulated through a software package and its performance has produced good results for the majority of digitising applications.

1.7.2 The Performance of a Digitising System

A digitising system's overall performance is the combination of several factors, each differing in performance, but taken collectively they offer the final result and can be categorised as follows (in no particular order):

touch-trigger probe and stylus performance
the CNC system performance
machine tool scale system
digitising performance

Touch Trigger Probe and Stylus Performance Factors

The definition of probe performance can be said to be "the repeatability of switching in a given direction", or to put this into perspective, this particular probe has an omnidirectional two sigma repeatability of just one micrometre. The basic contribution to the overall error performance is small, with the effect of the probe switching characteristics on accuracy being illustrated in Fig. 1.32d. There is some deflection of the probe stylus which occurs prior to the trigger signal being activated and this pre-travel is comprised of both the stylus bending and displacement – although it is repeatable in all directions to within one micrometre (i e two sigma) and does not vary with the direction of displacement. Owing to unknown workpiece surface characteristics, it is

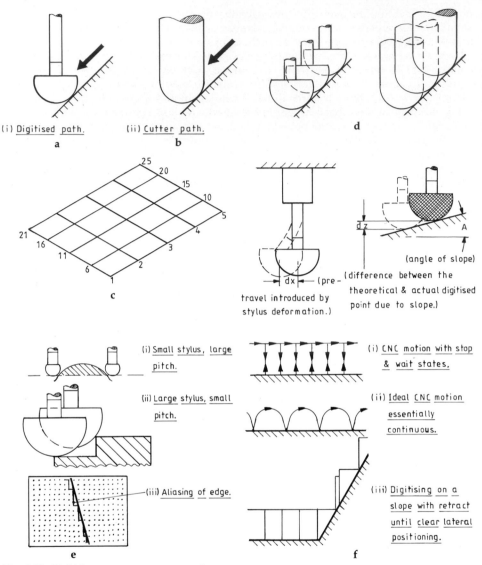

Fig. 1.32. Digitising a component on a machining centre – an overview of the operating principle. **a** The basic principle. **b** Consecutive points are used to form a cutter path. **c** The basic grid style and the order of digitising. **d** Causes of error by deformation and slope. **e** The effect of stylus and pitch variation. **f** Digitising motions. [Courtesy of Renishaw Metrology MAE.]

not possible to know the exact contact point between the stylus "ball" and the workpiece (Fig. 1.32d). Since the styli are not infinitely stiff, a small amount of bending occurs on contact with a part and this movement is termed pre-travel and occurs prior to the switching of the probe. The pre-travel variation cannot be accommodated by the probe's calibration and is shown as an error of surface duplication. In order to minimise the stylus pre-travel – due to bending variations – a range of stiff ceramic stemmed styli matching the diameters of commercially available ball-nosed endmills are available, with matching software. As an example of stylus bending of the stem, typical variations for a 50 mm long stylus are 0.038 0.050 mm.

As can be seen in Fig. 1.32d, the effect of pre-travel is to modify the theoretical point of contact. If "A" is the slope angle at the point of contact and "dx" is the pre-travel in the X–Y plane, the Z-plane value given will be lower by: dx/cos A. The solution is to use stiffer styli, and the shafts of 67% larger diameter than those previously used ensure that surface distortions are kept to a minimum.

The CNC System Performance

Any CNC system deals with the probe's signal in different ways and in real-time gauging, the controller can operate fast enough so that no delays in signal processing and recording the positional data occur. However, not all systems can be operated in real-time and errors through reading delays may occur. Such processing errors can be due to the sample time of the CNC; this is because these CNC systems scan external signals serially. Quite simply, the controller is not looking at the time a signal is given and as a result, the system does not respond until the signal has been "seen".

Assuming the worst case, the digitising has the following CNC error:

$$e = \frac{f \cdot t_s}{60}$$

where:
e = error in μm due to sample time
f = feedrate of CNC at the time of signalling in mm/min
t_s = sample time of input in ms
60 = a-units coefficient

For a typical CNC, the values will be:

$$t_s = 6\,ms$$
$$f - 100\,mm/min$$

Thus: $e = \dfrac{6 \times 100}{60} = 10\,\mu m$

The interpretation of this value is that there is a maximum error value of 10 μm owing to the sample time of the CNC and as this is a maximum value, the mean will be lower and closer to 5 μm which will affect the system resolution.

Normally, there should not be any other CNC performance factors associated with data capture, but other errors exist in the controller's system.

The final cutter path is a series of points to be linked by linear interpolation. To provide smooth motion, a CNC system buffers data and instigates the execution of a block of data just before the completion of the previous block. The zone in which this overlap occurs is dependent on the servo response of the machine tool – a heavy machine needs more time to change direction than a lighter one. Similarly, if the positions given by the program to the CNC system are very close, then in certain cases it may be seen that the final cutter path will deviate from the program path with less linearity than expected.

Machine Tool Scale System

Digitising techniques depend on the accuracy of the machine tool's positional measuring system. If this system is subject to errors then the digitising data will carry those errors.

Touch-trigger digitising is a relatively light-duty cycle for any machine tool. The greatest errors sourced in the scale system will be those due to changes in the ambient conditions during digitising – thermal effects within the machine tool. An encoder system mounted on the leadscrew itself – indirect feedback – will suffer directly from these changes, modifying the accuracy as a result, although the direct feedback linear scale systems are less affected by thermal growth problems. With the changes in response to market requirements, CNC system operation will also change and this will require the speed of digitising to increase. As the increased duty cycle of the servo systems increases, this promotes extra input into the machine tool and further distortions may be present within the encoder systems. Such increased heat input might cause the machine tool's geometry in addition to its linear accuracy to be affected.

Digitising Performances

For digitising techniques in two-dimensional operations, they operate at discrete intervals, with the probe being driven to a point in free space above its target and descending to the model's surface – this occurs irrespective of the part's slope, or geometry. Therefore there is a relationship between the original model and its digitised replica, which is dependent on the grid size and its pitch (see Fig. 1.32c). In the case of Fig. 1.32e(i), a large pitch is being used in conjunction with a small stylus. In such a case the surface definition will be lost, as the data cannot "see" between the discrete points on the grid. However, in Fig. 1.32e(ii), a larger stylus and a small pitch occur with this smaller pitch preventing overcutting, but the larger stylus diameter being unable to "see" small radii – in this condition a "metal-on" situation results.

Yet another digitising error occurs when the result of slopes lies within grids at indeterminate angles, as illustrated in Fig. 1.32e(iii). If a slope lies across the points of a mesh and meets along every point at this angle, such as 45° on a grid with pitches "x" and "y" being the same, then the line would be accurately defined. However, if the slope of such a line falls outside these parameters then "aliasing" will occur, such as the "saw-toothed" effect (Fig. 1.32e(iii)) produced instead of a line. Under these circumstances the solution is to reduce the grid size, increasing the chance of the line falling onto grid intersection points.

Model Stylus Considerations

If we appreciate that any digitising takes place within a predefined space and it can be applied to any model which may be placed within that space, with only certain limitations in the model's profile, this infers that there should be no re-entrant angles, or surface reversals which might prevent part removal from a mould.

Using touch-trigger probes the forces generated are light, being considerably less than those associated with the older technique of electronic tracing. Typical lateral forces are 50–100 g using a stylus with length of 50 mm, which can rise to 400–500 g in the vertical direction and, as a result, models should be rigid enough in structural integrity and the mounting to withstand these light loads.

An Efficient Digitising System

Clearly, digitising requires the system to store large quantities of data, and hard discs allow efficient mass storage in personal computers – meaning that the CNC system is capable of transmitting data whilst the part is digitised. Equally, a machining program will be generated from the same points and containing a number of passes; the program will be large – normally greater than that of most machine tool memories. It is anticipated that the controller should be able to receive the digitised program via an external source by "trickle feeding", on demand from the CNC. It is important that there is a provision for a suitable and efficient data communications channel which is accessible whilst the system is running.

Possibly the major consideration in digitising performance is the digitising speed, with the actual performance being controlled by the ability of the CNC to provide fast and continuous motion even when there are significant changes in direction. In Fig. 1.32f, some characteristics of digitising which are significant are illustrated. The motion of the machine from point to point should be continuous, without the delays associated with block read time and in position checking. The digitising sequence with "stop and think" states at the end of each motion can be seen in Fig. 1.32f(i); with this method, an average digitising time of 2.5 s is achieved. The ideal path occurs in Fig. 1.32f(ii), where the minimum of delays are present, offering significant reductions in processing time – about 1 s per point. Obviously the probe cannot contact the model prior to taking a reading, with the software noting the collisions in the X–Y plane before taking a reading in the Z plane. The final diagram showing digitising motions (Fig. 1.32f(iii)) highlights how the probe retracts as it collides with the model's surface before meeting its X–Y coordinate targets; with efficient processing this reduces the delay that might otherwise occur resulting from "stop and think" decision making.

The Development of Digitising for Mould Work

Several factors are predominant when digitising moulds on CNC machines. These factors relate to how data is used by the personal computers and software, so by utilising a computer effectively the database can be manipulated. For example, what was inconceivable when tracing moulds with hydraulic copying techniques, are now straightforward operations. With computer software, the data can be adjusted allowing for shrinkage factors which can be altered separately for each axis, with the computer insensitive to whether the shrinkage factor is 1% or 50%; however, families of parts based upon dimensional scaling are possible. Yet another advantage of using software is the "mirror-imaging" capability, which complements the scaling function; this "scaling" can produce "families" of related parts. As all these functions are common to one database, the digitising operation need only be performed once with the computing creating the necessary "data model" from which the mould is cut.

The operating principle for digitising, highlighted in Fig. 1.32, is shown practically in Fig. 1.33 on a vertical machining centre. This system is $2\frac{1}{2}$-dimensional rather than 3-D and to produce male/female mould transforms using $2\frac{1}{2}$-D geometrically is extremely difficult. However, the solution to this desirable capability is found by investigating linear data transformation techniques and a mathematical routine gives the desired transforms within practical limits.

This completes our review of digitising, with just one technique of digitising discussed amongst the variety of methods currently available. Digitised models offer a realistic means of "reverse engineering" which can offer competitive production

a

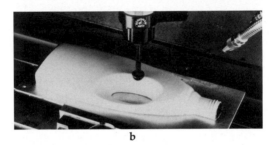

b

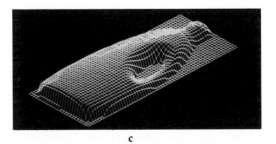

c

d

advantages, when companies are faced with either one-off or small batches. In the final section of this chapter a brief excursion into CAD/CAM will be presented, illustrating the benefits that accrue from its implementation.

1.8 Computer-aided Design and Manufacture

If companies decide upon the feasibility of purchasing CAD/CAM systems, inevitably they are governed by the software developed which determines the hardware that can be adopted. This software will have been written to run on a specific range of computers, although one might have some control over the available options for a particular computer. Prior to a discussion about the various aspects when choosing a computer suitable for a company's needs, there are several general questions which need consideration. A CAD/CAM system must be continuously developed, with improved software becoming available periodically. Therefore the software supplier must ensure that future "upgrades" will be available on the hardware for a realistic period of time. Certainly, a company having purchased the CAD/CAM system will rely heavily upon it and in the event of a breakdown, a hardware maintenance service that is readily available is essential.

Any calculations performed by the computer occur via the central processing unit, or CPU, with processor speed being measured in millions of instructions per second, or MIPs. The speed at which the processor achieves its calculations is an important factor, which determines how fast the applications software performs its task, although this is by no means the only factor requiring consideration. The applications software speed depends on how well it utilises the hardware capabilities, graphics software and operating system software. Furthermore, it will also depend on the memory available, together with the hard disk's access speed. With this in mind, the only way of obtaining a realistic comparison between computers is to run a "benchmark test", which typifies the work expected to be performed by the CAD/CAM in-service. Typically, the creation of toolpaths when machining a complex three-dimensional surface might suffice (see Fig. 1.34d).

Whenever we attempt an exercise in mental arithmetic, we hold the numerical values in our memory whilst performing the calculation. We also know how to achieve these calculations. A computer memory, or random access memory (RAM) has similarities to our mental capabilities, but it can retain significantly greater numbers in its memory. The expression used to determine a computer's memory is "megabytes", or millions of characters and by way of illustration, the *Concise Oxford Dictionary* might occupy about 5 megabytes of memory. The amount of memory

Fig. 1.33. Digitising techniques for milling dies and moulds. **a** A low cost digitising package. It allows the manufacture of dies and moulds from an original sample: a "digital replica". **b** A touch-trigger probe is moved to pre-determined points, e.g. in the X and Y plane, taking readings in the Z-axis. Probing density is programmable, each reading taking typically 2.5 s. The resulting data is transmitted to the on-line computer. **c** Once digitised, the data can be examined using wire frame graphics, either in part section, or section, or whole. Points can be reviewed, edited and redundant data eliminated. The data transform can be made on simple command, e.g. mirror image, scaling in 1, 2 or 3 axes, male/female transform. **d** The machining centre is set up with the required size cutter to make the replica. High accuracy means that little hand finishing is necessary. [Courtesy of Renishaw Metrology MAE.]

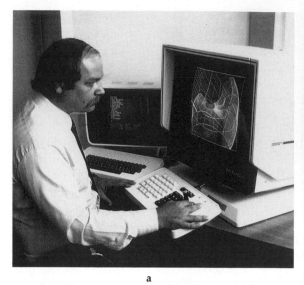

a

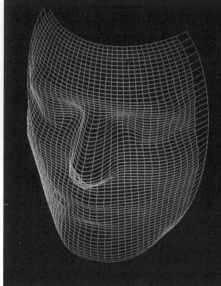

b

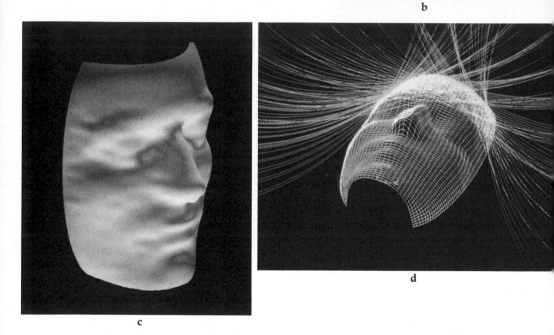

c

d

Fig. 1.34. The approximate stages in machining "sculptured surfaces". **a** A twin-screen CAD/CAM workstation with the sophisticated software necessary for 3-dimensional applications. **b** Isoparametric curves plotted for two different parameterisations of the same patch – the "Coon's patch", often termed a "wire-frame model". **c** Continuous tone picture using Painter's algorithm – illumination computed; "Lambert's law model". **d** Tool path graphics, illustrating cutter motions superimposed on a "wire-frame model". **e** Machined profile, using a 5-axis machining centre. [Courtesy of McDonnell Douglas/Boston Digital Corporation.]

e

Fig. 1.34. Continued

required is variable and will depend on the computer and the software it is currently running. Generally, we can say that the greater the memory capacity, the faster it runs – up to a point – as the "law of diminishing returns" applies and one must decide whether increased expenditure is worth the performance improvements.

A computer system must have adequate hard disk capacity which is used to store the CAD/CAM software and, additionally, to store the data created by the user, typically geometry files, or tape files to be sent to a machine tool via DNC, or similar. The important question for the user is not how much total hard disk capacity there is, but how much user space is available, as this depends on the area occupied by both the operating system and applications software. Naturally, it would be obvious to assume that the more disk space the better, but any data on the disk must be managed carefully to ensure it does not become cluttered with "old files". An important safeguard is to obtain "back-up files" kept separately either on a floppy disk, or tape, as hard disk failures can arise occasionally and lost data could prove disastrous to the company. It is usual to keep these "back-ups" in a fire-proof safe, ideally in a separate building to the CAD/CAM system. When a user only has sufficient disk space for the present needs, rather than for long term storage, then regular "back-ups" will allow tape files to be kept in order.

Clearly, it is necessary to discuss computer hardware requirements in conjunction with the software it is to run, and, to use a musical analogy, if the hardware is a violin then the software is the musical score, neither is of any use without the other. CAD/CAM software may be considered in three categories:

operating system software
graphics software
applications software

Let us briefly review these software applications in turn, considering probably the most important aspect – the operating system software – first. The software comprising the operating system controls the routine functions of the computer and as such influences how efficiently the computer performs its task. Typical operating systems such as UNIX and VMS tend to be more complex then MS-DOS. For example, both UNIX and VMS can perform several operations simultaneously, typically running the CAD/CAM system, post-processing and sending files to machine tools – usually termed "multi-tasking". The graphics software creates screen images, shaded pictures (Fig. 1.34c), line drawings and pop-up menus. Lastly, the applications software is the actual CAD/CAM software together with any associated software for either post-processors, DNC links, etc.

The quality of the screen graphics is an important feature for any visual representations such as CAD/CAM systems and one describes this aspect in terms of the maximum resolution of the monitor. A typical high-resolution monitor with a 19 inch screen has 1024×864 pixels (picture elements). Hence, the greater the number of pixels per inch of screen area, the better the picture resolution. Obviously, in order to obtain the monitor's maximum resolution, the computer and its software must be able to support such capability.

Earlier, we talked about the need to keep a "back-up" copy and just as important is the ability to accept data from other systems. Data can be stored either on $3\frac{1}{2}$ inch or $5\frac{1}{4}$ inch floppy disks, or, alternatively, via tape cartridges, or cassettes. Floppy disks can only hold up to 1.5 megabytes of memory, whereas tapes can conveniently hold large amounts of data. A limitation of tape storage is that the data can only be accessed sequentially and tapes tend to be large and unwieldly, with long programs taking time to read. Conversely, floppy disks allow quick access to files and at any point on the disk. A company's customers may have their own CAD/CAM facilities, so compatibility of each other's systems might figure as a high priority when the company considers its hardware requirements – particularly if both need to access each other's files.

Peripheral devices such as plotters and printers are obviously desirable elements in any CAD/CAM system, together with the need to "down-load" programs to the machine tools. Sufficient communication ports to which these peripheral devices can be linked is essential. Long CNC data files are often "down-loaded" to the machine tool's controller by a variety of means, but typically in a "block-to-block" fashion – particularly whenever enough "buffer storage" is not available to hold the complete program: termed "drip-feeding". In order to achieve uncorrupted data transfer of programs from one hardware device to another, efficient and compatible "handshaking" is essential, but more will be said on communications in chapter 2, Volume I.

Finally, once the company has decided which CAD/CAM system they feel will offer them the best compromise in the initial stages of implementation, they must bear in mind that at some later stage they may require to up-grade the system as their needs change. It is important that any system expansion can take place without a complete and costly overhaul of this system. Therefore, if it is envisaged that later up-grading is likely, this point should be addressed at an early stage of any feasibility assessment of prospective CAD/CAM systems.

1.8.1 Choosing Computerware – the Process of Elimination

As with many things in life, there is inevitably no clear-cut answer when one decides to purchase either hardware or software for a CAD/CAM system, as any final choice may depend upon many factors. Inevitably, the decisions taken about both hardware and software are often interrelated, with the suitability of one being dependent on the efficiency of the other. By the well-tried process of elimination, one can compromise to find a particular system which best suits a company's work, business set-up and staff. Generally, it is necessary to examine six areas when coming to a decision, namely:

functionality
software
applications
software
support
supplier

"Functionality" – computer jargon to explain its difference from "function" (which has other connotations in the computer world) – clearly refers to the purpose of the system. For example, does one require a computer-aided draughting and design (CADD) system having a 3-dimensional modelling capability, or conversely, would a 2-dimensional draughting system be adequate? Obviously, the system chosen limits the range of appropriate software. "User-friendliness", compatibility of suppliers'/customers' systems, trained personnel available, proposed future system developments and a company's business requirements should all be investigated when deciding the best software system for the company.

A typical "application" factor might include whether a system is for a "single-user", namely for one terminal, or if it is to be for "multi-use" such as interfacing with other computers. Therefore, will one need either a central database, or perhaps terminals at other locations? This means that application considerations in conjunction with the software chosen for a company's needs will influence the hardware purchased.

A major, but often underrated, consideration should be "support", as even the most suitable system can be a liability when poorly installed, or not accompanied by software up-dates and quick repairs to hardware. Yet another "support" area which should be considered is the relative practical training of a vendor's staff, as if this is not effectively given, together with appropriate documentation, then this can influence the CAD/CAM's impact within the company at large.

The choice of supplier will be made simpler by highlighting these critical factors as only a few suppliers will meet the company's objectives. Often the computer firms short-listed will have a specific knowledge of a company's manufacturing enterprise and their previous experience will be invaluable, as this saves both time and expense. Often such CAD/CAM suppliers know more about the scope of this equipment for a specific company's applications and can advise them accordingly and, in this way, obtain the most cost-effective equipment. This is not always the case, however, as the choice of software may lead to only a single supplier – possibly with a limited hardware option. If the company purchasing the CAD/CAM system is a "first-time user" they should act with caution, as the temptation might be to obtain a "limited" system owing to its lower cost.

Sometimes corporate aspirations can be important at the initial feasibility stage in the purchase of a CAD/CAM system, because both the scope and function of such equipment can influence its effectiveness within the company. It is possible to implement such systems gradually within a company and it is worth considering obtaining the advice of an experienced consultant in the field. Such people could map out a strategic program of the functional implementation of a computer system over the complete spectrum of a company's operation in stages over a two-year time period. This should be considered the maximum time for the original CAD/CAM system, as the technology is changing so fast that any initial choice will be obsolete five years later.

Any initial implementation program to choose the software should simultaneously consider the potential links at future stages to process planning, manufacturing and metrological considerations. Such manufacturing philosophies will inevitably lead to choosing some form of multi-user system using either a mini-computer, or a workstation network – this latter approach is possibly more flexible, but here we must leave the subject, as it becomes highly "user-specific" to a company's needs.

Once a system has been chosen, whether it is a workstation network or a stand-alone system, there are several additional pieces of hardware available which improve and enhance the system still further. Occasionally a company's software needs will dictate that "extras" are required; some investigations into the options are necessary and are described below:

mathematical co-processor – this speeds up calculations

additional memory – i.e. random access memory (RAM)

extra storage facilities – e.g. disk

back-up and archiving options

high resolution colour monitors

input devices – tablet, mouse, etc.

output devices – plotter, printer, etc.

document management

historic data handling – scanning

security

Increasingly, choices in back-up and archiving techniques are becoming popular and as systems become larger, more data is created, so an important security feature must be the ability to provide back-up files speedily. As an example of this importance, we can consider that a network of six personal computers having produced a day's work necessitates an hour a day attended loading of files on floppy disk space, or half an hour unattended loading using a magnetic medium. Similarly, most cartridge tapes are limited to 40–60 Mb of data, although some can hold more, up to 300, or even 2300 Mb (i.e. 2.3 Gb). It can be appreciated that both the selection and implementation of any back-up and archive unit can seriously impede the productivity and security of a CAD/CAM system.

Once the system has been chosen it is possible to enhance the equipment and improve its efficiency still further by additional "refinements". Using twin screens (see Fig. 5.34a) improves user performance, by allowing one for detail and the other for the overall picture; a third screen can be employed for text only. If the second screen is just a "text only" version, it cannot cope with high-resolution presentation. Graphics monitors should always be coloured, as even very basic CAD/CAM software allows a palette of around sixteen colours, whereas more extensive colour capabilities can rarely be justified in engineering applications – some systems offer 256 discrete colour shades out of a potential palette of 64 million.

Further, the geometric design and draughting benefits – aesthetically allowing circles to appear as round and lines straight, from increased resolution capacities up to 1024 × 786 pixels (i.e. points of resolution displayed) offer a more realistic visual interpretation, but in general do not aid in their improved interpretation to any great extent. It should be pointed out that higher-resolution screens do help us appreciate surfaces (Fig. 1.34c), whereas schematics (Fig. 1.34b) do not justify such high resolution to gain a visual understanding of the displayed artefact. With the improvements in the effective use of "window" technology for most software CAD/CAM systems, the use of a multi-button mouse has largely superseded the menu tablet, which should be avoided if possible.

When selecting printers/plotters, one should aim for the level of print quality suitable for a company's particular text presentations and a plotter capable of coping with the maximum size of paper used. Plotter speed varies with cost and this becomes an important consideration as more operators use the system.

Document management is a relatively recent concept with most CAD/CAM systems with the advent of quality assurance procedures becoming a dominant factor in most trading companies of manufactured goods. Any CAD/CAM system of late must conform to the quality assurance system of presentation, EN29000 (previously BS5750 part 2), but this can be open to a degree of interpretation by the software companies with reference to what level of detail is required and what security options to adopt, together with techniques in retrieving them.

Finally, the skill in choosing any "computerware" for CAD/CAM applications lies in selecting both software and hardware which gives the best facilities at present and the greatest scope in the future, whilst obviously having the optimum value for money. Therefore, it is essential to select the system according to a company's present and specific business needs and their future plans.

In the final section of this chapter we will briefly consider both the problems of representing sculptured surfaces on screens and just some of the techniques used to machine them.

1.8.2 Sculptured Surfaces and their Machining Problems

In CAD/CAM systems, curves and subsequent surfaces result in two functional demands of the system:

curve fitting
curve design

In "curve fitting" the fitting of a curve, or surface, is called for through a set of defined points having a smooth transition from one point to the next. "Curve design" entails the modification of the curve equation parameters either directly or indirectly, to observe what shape may be developed.

Most sophisticated 3-dimensional CAD/CAM systems have "curve fitting" capabilities, often using a modified cubic equation-based technique. Such methods mean that in the equation defining the curve there are individual sums – including cubed factors – which, when added to squared factors and more sums them finally to an individual number, representing the curve mathematically as exemplified below:

$$U^3 a_3 + U^2 a_2 + Ua + a_0 = r$$

As each equation is developed, this produces a discrete curve which can be more easily defined with respect to its start and end points, in addition to the curve's slope at each point.

This technique of "curve fitting" is not new and such indirect methods were devised in the early 1960s, making it relatively easy to manipulate these curves – without recourse to modifying the different equation parameters. In a typical system, a complicated curve would be comprised of several discrete curves – termed "spline" – whereas a surface is simply a curve with an extra dimension. For a "curve fitting" the cubic method is particularly suited, although a modified cubic approach that can accommodate the uneven spacing of "nodes" – the curve start and end points – has particular benefits when digitising surfaces.

It was a Frenchman, Bezier, who, whilst working for Renault was intrigued by car body design and found the "point and slope" technique rather inconvenient for curve design. His philosophy was to find a way of manipulating the individual parameters contained within the basic equation in an easier manner also using the indirect method. Bezier used an open polygon – a plane figure of many angles and straight sides by which a curve approximating to it passes through the start and end point of the open polygon: resulting in a designer changing the polygon and, as such, achieving different results. Having more defined points in the polygon gives more flexible control for surface manipulation; furthermore, the curves generated are formed by equations comprised of parameters raised to higher powers than the cubic varieties, having longer and more complex mathematical expressions. Such a curve is a discrete segment in a complex curve and these segments must be joined together.

With the Bezier technique the transition between curve segments, or patches – the surface equivalent to a line segment – requires close study by the designer. A further refinement not developed by Bezier, but incorporating Bezier mathematics, was the "B-Splines", which ensure a smooth transition between segments/patches. Yet another, later, development was the non-uniform "B-Splines" which catered for the uneven spacing of nodes.

Terminology which is not very common but is associated with the term "NURBS" includes the rational and non-rational parametric surfaces, which we will define shortly. Returning to the rational parametric surface, this may be represented in many forms with mathematical precision. The cubic non-rational variety cannot express a 90° arc with mathematical precision, although it has adequate accuracy for any machining requirements. Most CAD/CAM systems have a variety of other software techniques to define such elements as circles, spheres and cylinders with the desired precision. However, rational parametrics have a single bias for form generation, which provides a tidy programming solution by eliminating different pieces of software, although whether this helps or hinders the user is an open debate.

This brings us back to "NURBS", referred to earlier, which is the amalgamation of rational parametric surfaces together with non-uniform B-Splines resulting in Non-Uniform Rational B-Splines – "NURBS".

Whenever a data file is transferred between two CAD/CAM systems having different surface definition methods, including equations with one incorporating parameters higher than cube orders, the receiving CAD/CAM system can break down the surfaces into smaller patches and as such they can be redefined using cubic form. Any differences that might occur through this redefinition of the mathematical expressions are so small as to be insignificant, for the practical purposes of machining contour. It should be said that nearly all today's systems have graphical interfaces, with the actual curve/surface profile manipulations being indirect, but the underlying theories and attributes still apply.

It is not possible in the space provided to discuss the process of surface construction and software manipulation for all CAD/CAM systems available. Even on just one system there are a variety of techniques that can be utilised to achieve similar machining results. As an example, we will consider one problem that may occur in the machining of a bottle mould cavity (Fig. 1.33d), at the portion where a "gouge" situation could be present at the transition between the main body and the neck. Assuming that a CAD/CAM system has "multi-surface" capability offering "gouge" avoidance" of the cutter, the complete cavity might be machined successfully employing this facility – although this would be time consuming, owing to the complex computations required looking for "gouge" situations. Possibly the best solution in this case might be to machine the main body and neck without "gouge avoidance", then apply it to a small selection of the neck/body transition. In order to reduce the user input time for "multi-surface" machining operations through many "keystrokes", "macros" have been developed – capturing a sequence of "keystrokes" in a simple command, in conjunction with APT (i.e. Automatically Programmed Tool) techniques.

So far we have briskly described the philosophy and mathematics used to develop sculptured surfaces which are becoming an increasingly important feature in manufacturing today. Typical processes requiring sculptured and smooth surfaces are die cavities, aerofoil sections, impeller blades, foundry patterns, body-panel pressings, etc. Inevitably, this means that in order to produce such free-form shapes a CNC milling technique should be adopted, in order to minimise tooling costs. Obviously if the part is to be manufactured on the machining centre it is necessary to produce a path for the cutter to follow (see Fig. 1.35) although the problems in determining the strategy when machining such sculptured surfaces are not simply restricted to the manufacturing operation alone. In order to verify the desired shape for our part, it is often prudent to cut a model from either remeltable wax, or foam, but this becomes impractical for large components.

When machining any 3-dimensional/sculptured surface, this is achieved by line milling along the contour, with the precise adaption of the tool geometry around the contoured profile being dependent upon both the tool geometry and its guidance. The tool's path can be evaluated in terms of the scallop width and height, whilst the tool guidance mode might be established using model scanning/digitising (see section 1.7.1).

If three-axes milling is used for machining contours, it means that we can only guide the tip of the tool within our "working envelope", as the tool's axis is maintained in a constant relationship to the profile. Often ball-ended milling cutters are used, or alternatively, either barrel-shaped cutters or indeed "fully-defined" APT cutters may be utilised. However, if we confine our comments to the ball-ended cutters, it is possible to use three different cutting modes when three-axes milling:

plunge cutting
constant-depth cutting
reverse cutting

The tool engagement modes will vary depending upon the curvature of the surface shape. The fixed cutter axis in three-axes milling operations produces unfavourable cutting conditions, whereas favourable tool geometry adaptions are possible with respect to the workpiece contours using five-axes milling techniques. As such, five-axes milling implies simultaneous technological advantages through control of both the tool's cutting point and the cutter's axis. Furthermore, five-axes milling represents the general case with respect to the milling process, as there are no additional restrictions that can influence the potential degrees of freedom for the rotating cutter. This infers that the cutter's orientation with respect to the workpiece contour offers favourable cutting conditions, and along many sculptured surfaces an exact relation-

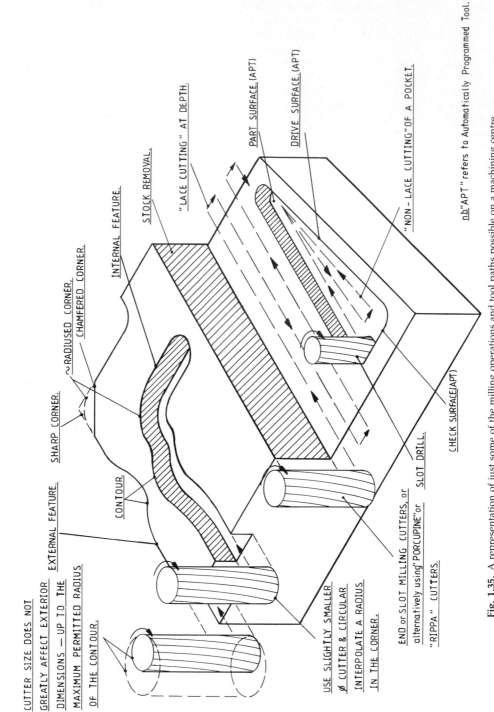

CUTTER SIZE DOES NOT
GREATLY AFFECT EXTERIOR
DIMENSIONS — UP TO THE
MAXIMUM PERMITTED RADIUS
OF THE CONTOUR.

EXTERNAL FEATURE.

SHARP CORNER.

~RADIUSED CORNER.
CHAMFERED CORNER.

INTERNAL FEATURE.

STOCK REMOVAL.

"LACE CUTTING" AT DEPTH.

PART SURFACE.(APT)

DRIVE SURFACE.(APT)

"NON-LACE CUTTING"OF A POCKET.

nb."APT" refers to Automatically Programmed Tool.

CONTOUR.

USE SLIGHTLY SMALLER
Ø CUTTER & CIRCULAR
INTERPOLATE A RADIUS
IN THE CORNER.

END or SLOT MILLING CUTTERS, or
alternatively using"PORCUPINE"or   SLOT DRILL.
"RIPPA" CUTTERS.

CHECK SURFACE(APT)

Fig. 1.35. A representation of just some of the milling operations and tool paths possible on a machining centre.

Fig. 1.36. The complete machining of an impeller, by utilisation of a 5-axis horizontal machining centre in conjunction with a high-level programming language. The "sculptured surfaces" are produced with great form accuracy and at optimum metalcutting efficiency. [Courtesy of Scharmann Machine Ltd.]

ship is established. Utilising five-axes, or three-axes milling techniques for sculptured surfaces causes a scalloped surface topography, as shown in Fig. 1.36 for five-axes impeller milling. As one can appreciate from Fig. 1.36, a greater overlapping of the milled paths results using five-axes face milling whilst maintaining the same scallop height as for three-axes milled contours. As a consequence, there will be a reduction in the number of milled paths over the profile leading to a reduction in the time taken to machine the part.

By using CAD/CAM systems, a feature available with the more sophisticated software is the ability to machine the surface profile within prescribed workpiece tolerances automatically. Using five-axes allows the machining to occur with constant, or variable, "tilt angles". The "tilt angle" can be defined by the surface normal to the tool axis at the point of contact with the part's surface. Using variable "tilt angles" alleviates possible cutter collisions during feeding when normal to the contour of the part, whilst achieving the greatest permissible scallop width – for improved stock removal and optimum scallop/cusp height reductions, further aiding the effort in part finishing.

Not only does five-axes milling offer the ideal cutter/workpiece geometric relationship during sculptured surface manufacture, it improves the tool wear characteristics considerably, resulting in extended tool life over three-axes profiling methods and greater protection to the tool by minimising the forces on the cutter. However, whatever cutting strategy is employed, care must be taken to avoid the cutter "gouging" other features on the workpiece, typically when machining re-entrant angles on the part. Using APT, or NMG (i.e. Numerical Master Geometry) provides for the definition of "check surfaces" (see Fig. 1.35) which are surfaces beyond which the cutter's tool path cannot exceed – "gouging avoidance". Such techniques effectively provide a means of blending one contour (patch) to another, smoothly. This means that as a ball-nosed cutter cannot get into a corner between the surface being cut and the "check surface", a radius will be developed. Using the "check surface" facility means that it becomes much simpler to implement than to define the blend explicitly.

It would have been quite easy to expand our thoughts and ideas considerably in this chapter on virtually every topic covered, but it was not the intention to explain all of the problems and solutions to turning and machining centre programming. Indeed, it would have been totally impossible. Therefore a concise and generalised treatment has been given, simply to give the reader the essence of programming.

In chapter 2, Volume I, we will discuss the various merits of Flexible Manufacturing Cells and Systems and describe just some of the problems that must be overcome for their effective utilisation within a company. Lastly we will go on to consider the latest trend in turning/machining centres – "sub-micron" machining techniques, whilst explaining the technological problems that must be overcome to machine at such highly accurate levels of manufacture.

Appendix
National and International Machine Tool Standards

Determination of Accuracy and Repeatability of Positioning of CNC Machine Tools

Date of issue	Standard	Country of origin
1972	NMTBA	USA
1977	VDI/DGQ 3441	Germany
1985	BS 4656: PART 16	Great Britain
1987	BS 4656: PART 16 (AMENDED)	Great Britain
1988	ISO 230–2	International
1991	BS 3800: PART 2	Great Britain

BS3800: General Tests for Machine Tools

PART 1: 1990 Code of practice for testing geometric accuracy of machines operating under no load, or finishing operations;
geometric and practical test methods,
definitions,
use of checking instruments,
explanation of tolerances,
description of preliminary checking operations,
description of accuracy of instruments required.

PART 2: 1991 Statistical methods for determination of accuracy and repeatability of machine tool;
linear and rotary positioning errors applied to CNC machine tools,
angular (pitch, yaw and roll) and straightness positioning errors applied to CNC and manually controlled machine tools.

PART 3: 1990 Method of testing performance of machines operating under loaded conditions in respect of thermal distortions;
thermal distortion of structure,
thermal drift of axis drives.

Glossary of Terms

A AXIS – The axis of rotary motion of a machine tool member or slide about the X axis.

ABSOLUTE ACCURACY – Accuracy as measured from a reference which must be specified.

ABSOLUTE DIMENSION – A dimension expressed with respect to the initial zero point of a coordinate axis.

ABSOLUTE POINT (Robots) – Equivalent to absolute coordinates in NC machines. The coordinates of a data point are defined in relation to an absolute zero.

ABSOLUTE PROGRAMMING – Programming using words indicating absolute dimensions.

ABSOLUTE READOUT – A display of the true slide position as derived from the position commands within the control system.

ABSOLUTE SYSTEM – NC system in which all positional dimensions, both input and feedback, are measured from a fixed point of origin.

ACCANDEC – (Acceleration and deceleration) Acceleration and deceleration in feedrate; it provides smooth starts and stops when operating under NC and when changing from one feedrate value to another.

ACCEPTANCE TEST – A series of tests which evaluate the performance and capabilities of both software and hardware.

ACCESS TIME – The time interval between the instant at which information is: 1. called for from storage and the instant at which delivery is completed, i.e., the read time; 2. ready for storage and the instant at which storage is completed, i.e., the write time.

ACCUMULATOR – A part of the logical arithmetic unit for a computer. It may be used for intermediate storage to form algebraic sums, or for other intermediate operations.

ACCURACY – 1. Measured by the difference between the actual position of the machine slide and the position demanded. 2. Conformity of an indicated value to a true value, i.e., an actual or an accepted standard value. The accuracy of a control system is expressed as the deviation or difference between the ultimately controlled variable and its ideal value, usually in the steady state or at sampled instants.

ACTIVE CONTROL – A technique of automatically adjusting feeds and/or speeds to an optimum by sensing cutting conditions and acting upon them.

ACTIVE STORAGE – That part of the control logic which holds the information while it is being transformed into motion.

ADDRESS – A character or group of characters at the beginning of a word which identifies the data allowed in the word.

ADDRESS BLOCK FORMAT – A block format in which each word contains an address.

ALGOL – (Algorithmic Language) Language used to develop computer programs by algorithm.

ALGORITHM – A rule or procedure for solving a mathematical problem that frequently involves repetition of an operation.

ALPHANUMERIC or ALPHAMERIC – A system in which the characters used are letters A to Z, and numerals 0 to 9.

ALPHANUMERIC DISPLAY – Equipment, such as a CRT, which is capable of displaying only letters, digits and special characters.

AMPLIFIER – A signal gain device whose output is a function of its input.

AMPLITUDE – Term used to describe the magnitude of a simple wave or simple part of a complex. The largest or crest value measured from zero.

ANALOG – In NC the term applies to a system which utilises electrical voltage magnitudes or ratios to represent physical axis positions.

ANALOG DATA – The information content of an analog signal as conveyed by the value of magnitude of some characteristics of the signal such as the amplitude, phase, or frequency of a voltage, the amplitude or duration of a pulse, the angular position of a shaft, or the pressure of a fluid.

ANALOG SIGNALS – Physical variables (e.g., distance, rotation) represented by electrical signals.

ANALOG-TO-DIGITAL (A/D) CONVERTER – A device that changes physical motion or electrical voltage into digital factors.

AND – A logical operator which has the property such that if X and Y are two logic variables, then the function "X and Y" is defined by the following table:

X	Y	X and Y
0	0	0
0	1	0
1	0	0
1	1	1

The AND operator is usually represented in electrical rotation by a centred dot "·", and in FORTRAN programming notation by an asterisk "*" within a Boolean expression.

AND-GATE – A signal circuit with two or more inputs. The output produces a signal only if all inputs received coincident signals.

APPLICATION PROGRAMS – Computer programs designed and written to value a specific problem.

APT – (Automatically Programmed Tools) A universal computer-assisted program system for multi-axis contouring programming. APT III – Provides for five axes of machine tool motion.

ARC CLOCKWISE – An arc generated by the coordinated motion of two axes in which curvature of the path of the tool with respect to the workpiece is clockwise, when viewing the plane of motion from the positive ddirection of the perpendicular axis.

ARC COUNTERCLOCKWISE – (Substitute "Counterclockwise" for "Clockwise" in "Arc Clockwise" definition.)

ARCHITECTURE – Operating characteristics of a control system, or control unit, or computer.

ASCII – (American Standard Code for Information Interchange) A data transmission code which has been established as an American Standard by the American Standards Association. It is a code in which 7 bits are used to represent each character. (Also USASCII.)

ASSEMBLY – The fitting together of a number of parts to create a complete unit.

ASSEMBLY DRAWING – The drawing of a number of parts which shows how they fit together to construct a complete unit.

ASYNCHRONOUS – Without any regular time relationship.

ASYNCHRONOUS TRANSMISSION – The transmission of information in irregular sections, with the time interval of each transmission varying and each section being identified by a stop and start signal.

ATTRIBUTE – A quality that is characteristic of a subject.

AUTOMATED ASSEMBLY – The application of automation to assembly.

AUTOMATION – The technique of making a process or system automatic. Automatically controlled operation of an apparatus, process, or system, especially by electronic devices. In present day terminology, usually used in relation to a system whereby the electronic device controlling an apparatus or process also is interfaced to and communicates with a computer.

AUXILIARY FUNCTION – A function of a machine other than the control of the coordinates of a workpiece or cutter – usually on–off type operations.

AXIS – 1. A principal direction along which a movement of the tool or workpiece occurs. 2. One of the reference lines of a coordinate system.

AXIS (Robots) – A moving element of a robot or manipulator.

AXIS INHIBIT – Prevents movement of the selected slides with the power on.

AXIS INTERCHANGE – The capability of inputting the information concerning one axis into the storage of another axis.

AXIS INVERSION – The reversal of normal plus and minus values along an axis which makes possible the machining of a left-handed part from right-handed programming or vice-versa. Same as mirror image.

B

B AXIS – The axis of rotary motion of a machine tool member or slide about the Y axis.

BACKGROUND – In computing the execution of low priority work when higher priority work is not using the computer.

BACKGROUND PROCESSING – The automatic execution of computer programs in background.

BACKLASH – A relative movement between interacting mechanical parts, resulting from looseness.

BAND – The range of frequencies between two defined limits.

BASE – A number base. A quantity used implicitly to define some system of representing numbers by positional notation. Radix.

BATCH – A number of items being dealt with as a group.

BATCH PROCESSING – A manufacturing operation in which a specified quantity of material is subject to a series of treatment steps. Also, a mode of computer operations in which each program is completed before the next is started.

BAUD – A unit of signalling speed equal to the number of discrete conditions or signal events per second; 1 bit per second in a train of binary signals, and 3 bits per second in an octal train of signals.

BEHIND THE TAPE READER – A means of inputting data directly into a machine tool control unit from an external source connected behind the tape reader.

BENCHMARK – A standard example against which measurements may be made.

BILL OF MATERIALS – A listing of all the parts that constitute an assembled product.

BINARY – A numbering system based on 2. Only the digits 0 and 1 are used when written.

BINARY CIRCUIT – A circuit which operates in the manner of a switch, that is, it is either "on" or "off".

BINARY CODED DECIMAL (BCD) – A number code in which individual decimal digits are each represented by a group of binary digits; in the 8-4-2-1 BCD notation, each decimal digit is represented by a four-place binary number, weighted in sequence as 8, 4, 2 and 1.

BINARY DIGIT (BIT) – A character used to represent one of the two digits in the binary number system, and the basic unit of information or data storage in a two-state device.

BLOCK – A set of words, characters, digits, or other elements handled as a unit. On a punched tape, it consists of one or more characters or rows across the tape that collectively provide enough information for an operation. A "word" or group of words considered as a unit separated from other such units by an "end of block" character (EOB).

BLOCK DELETE – Permits selected blocks of tape to be ignored by the control system at discretion of the operator with permission of the programmer.

BLOCK DIAGRAM – A chart setting forth the particular sequence of operations to be performed for handling a particular application.

BLOCK FORMAT – The arrangement of the words, characters and data in a block.

BODE DIAGRAM – A plot of log amplitude ratio and phase angle as functions of log frequency, representing a transfer function.

BOOLEAN ALGEBRA – An algebra named after George Boole. This algebra is similar in form to ordinary algebra, but with classes, propositions, yes/no criteria, etc., for variables rather than numeric quantities, it includes the operator's AND, OR, NOT, EXCEPT, IF THEN.

BOOTSTRAP – A short sequence of instructions, which when entered into the computer's programmable memory will operate a device to load the programmable memory with a larger, more sophisticated program – usually a loader program.

BUFFER STORAGE – 1. A place for storing information in a control for anticipated transference to active storage. It enables control system to act immediately on stored information without waiting for the tape reader. 2. A register used for intermediate storage of information in the transfer sequence between the computer's accumulators and a peripheral device.

BUG – An error or mistake.

BULK MEMORY – A high capacity auxiliary data storage device such as a disk or drum.

BUS – A conductor used for transmitting signals or power between elements.

BYTE – A sequence of adjacent bits, usually less than a word, operated on as a unit.

C

C AXIS – The axis of rotary motion of a machine tool member or slide about the Z axis.

CALIBRATION – Adjustment of a device, such that the output is within a specified tolerance for particular values of the input.

CANCEL – A command which will discontinue any canned cycles or sequence commands.

CANNED CYCLE – A preset sequence of events initiated by a single NC command, e.g., G84 for NC tap cycle. Also fixed cycle.

CANONICAL FORM – A standard numerical representation of data.

CATHODE RAY TUBE (CRT) – A display device in which controlled electron beams are used to present alphanumeric or graphical data on a luminescent screen.

CENTRAL PROCESSING UNIT (CPU) – The portion of a computer system consisting of the arithmetic and control units and the working memory.

CHANNEL – A communication path.

CHARACTER – One of a set of symbols. The general term to include all symbols such as alphabetic letters, numerals, punctuation marks, mathematic operators, etc. Also, the coded representation of such symbols.

CHIP – A single piece of silicon which has been cut from a slice by scribing and breaking. It can contain one or more circuits but is packaged as a unit.

CIRCULAR INTERPOLATION – 1. Capability of generating up to 90 degrees of arc using one block of information as defined by EIA. 2. A mode of contouring control which uses the information contained in a single block to produce an arc of a circle.

CLDATA – Cutter location data (see CLFILE).

CLEAR – To erase the contents of a storage device by replacing the contents with blanks or zeros.

CLEARANCE DISTANCE – The distance between the tool and the workpiece when the change is made from rapid approach to feed movement to avoid tool breakage.

CLFILE – Cutter location file (see CLDATA).

CLOCK – A device which generates periodic synchronisation signals.

CLOSED LOOP – A signal path in which outputs are fed back for comparison with desired values to regulate system behaviour.

CNC – Computer (Computerised) Numerical Control – A numerical control system wherein a dedicated, stored program computer is used to perform some or all of the basic numerical control functions.

COMMAND – An operative order which initiates a movement or a function.

COMPATIBILITY – The interchangeability of items.

COMPILER – A program which translates from high-level problem-oriented computer languages to machine-oriented instructions.

COMPONENT – One of the parts of which an entity is composed.

COMPUTER – A device capable of accepting information in the form of signals or symbols, performing prescribed operations on the information, and providing results as outputs.

COMPUTER-AIDED DESIGN (CAD) – A process which uses a computer in the creation or modification of a design.

COMPUTER-AIDED DESIGN/COMPUTER-AIDED MANUFACTURE (CADCAM) – The integration of computer-aided design with computer-aided manufacture.

COMPUTER-AIDED ENGINEERING (CAE) – The use of computing facilities in the integration of all aspects of design and manufacture to create an integrated engineering facility.

COMPUTER-AIDED MANUFACTURE (CAM) – A process which uses a computer in the management, control or operation of a manufacturing facility.

COMPUTER PART PROGRAMMING – The preparation of a part program to obtain a machine program using a computer and appropriate processor and part processor.

CONFIGURATION – The manner in which items are arranged.

CONTINUOUS PATH OPERATION – An operation in which rate and direction of relative movement of machine members is under continuous numerical control. There is no pause for data reading.

CONTOURING – An operation in which simultaneous control of more than one axis is accomplished.

CONTOURING CONTROL SYSTEM – An NC system for controlling a machine (milling, drafting, etc.) in a path resulting from the coordinated, simultaneous motion of 2 or more axes.

CONTROLLED PATH (Robots) – The straight line motion of a defined offset tool point between programmed points. All robot axes are interpolated through the programmed span.

CONTROL TAPE – A tape on which a machine program is recorded.

COORDINATE DIMENSIONING – A system of dimensioning based on a common starting point.

COORDINATE DIMENSIONING WORD – 1. A word in a block of machining information that provides instruction for one of the machine's axes. 2. A word defining an absolute dimension.

CORE MEMORY – A high-speed random access data storage device utilising arrays of magnetic ferrite cores, usually employed as a working computer memory.

CORE RESIDENT – Pivotal programs permanently stored in core memory for frequent execution.

COUNTER – A device or memory location whose value or contents can be incremented or decremented in response to an input signal.

CURSOR – Visual movable pointer used on a CRT by an operator to indicate where corrections or additions are to be made.

CUTTER DIAMETER COMPENSATION – A system in which the programmed path may be altered to allow for the difference between actual and programmed cutter diameters.

CUTTER OFFSET – 1. The distance from the part surface to the axial centre of a cutter. 2. An NC feature which allows an operator to use an oversized or undersized cutter.

CUTTER PATH – The path described by the centre of a cutter.

CYCLE – 1. A sequence of operations that is repeated regularly. 2. The time it takes for one such sequence to occur.

CYCLE TIME – The period required for a complete action. In particular, the interval required for a read and a write operation in working memory, usually taken as a measure of computer speed.

CYCLING CONTROL – A fundamental level machine control which programs the machine through dial or plugboard input.

D

DAMPING – A characteristic built into electrical circuits and mechanical systems to prevent rapid or excessive corrections which might lead to instability or oscillatory conditions.

DATA – Facts or information prepared for processing by, or issued by, a computer.

DATABASE – Comprehensive files of information having a specific structure such that they are suitable for communication, interpretation and processing by both human and automatic means.

DATA POINT – A programmed point which contains tool plant coordinate data and functional information.

DEAD BAND – The range through which an input can be varied without initiating response, usually expressed in percentage of span.

DEAD TIME – The interval between initiation of a stimulus change and the start of the resulting response.

DEAD ZONE – A range of inputs for which no change in output occurs.

DEBUG – To detect, locate, and remove mistakes from computer software or hardware.

DECADE – A group of assembly of ten units.

DECADE SWITCHING – Use of a series of switches each with ten positions with values of 0 to 9, in which adjacent switches have a ratio of value of 10:1.

DECIMAL CODE – A code in which each allowable position has one of 10 possible states. (The conventional decimal number system is a decimal code.)

DECODER – A circuit arrangement which receives and converts digital information from one form to another.

DEDICATED – Devoted to a particular function or purpose.

DEVIATION – The error or difference between the instantaneous value of the controlled variable and the setpoint.

DIAGNOSTIC ROUTINE – A program which locates malfunctions in hardware or software.

DIGITAL – Representation of data in discrete or numerical form.

DIGITAL COMPUTER – A computer that operates on symbols representing data, by performing arithmetic and logic operations.

DIGITAL-TO-ANALOG (D-A) CONVERSION – Production of an analog signal, whose instantaneous magnitude is proportional to the value of a digital input.

DIGITISE – To obtain the digital representation of a measured quantity or continuous signal.

DIRECTOR – A term used to designate an NC control unit.

DISCRETE – State of being separate or distinct, as opposed to a continuously varying state or condition.

DISCRETE COMPONENT CIRCUIT – An electrical circuit, implemented with individual transistors, resistors, diodes, capacitors, or other components.

DISK – A device on which information is stored.

DISK MEMORY – A non-programmable, bulk storage, random access memory consisting of a magnetisable coating on one or both sides of a rotating thin circular plate.

DISPLAY – Lights, annunciators, numerical indicators, or other operator output devices at consoles or remote stations.

DISTRIBUTED COMPUTER NETWORK – A collection of computers which can communicate with each other.

DISTRIBUTED PROCESSING – The processing of information on a distributed computer network in such a manner as to improve the overall efficiency of the task.

DITHER – An electrical oscillatory signal of low amplitude and of a predetermined frequency imparted to a servo valve to keep the spool from sticking.

DNC – (Direct Numerical Control) Numerical control of machining or processing by a computer.

DOCUMENTATION – The group of techniques necessarily used to organise, present, and communicate recorded specialised knowledge.

DOUBLE PRECISION – The use of two computer words to represent a number.

DOWNTIME – The interval during which a device is inoperative.

DRIFT – An undesired change in output over a period of time, which is unrelated to input, operating conditions, or load.

DRIVER – A program or routine that controls external peripheral devices or executes other programs.

DUMP – To copy the present contents of a memory onto a printout or auxiliary storage.

DWELL – A timed delay of programmed or established duration, not cyclic or sequential, i.e., not an interlock or hold.

DYNAMIC GAIN – The magnitude ratio of a steady-state output to a sinusoidal input signal.

E

EBCDIC – Extended binary coded decimal interchange code.

EDIT – To modify a program, or alter stored data prior to output.

EDITOR – A computer program which provides the ability to edit.

EIA STANDARD CODE – Any one of the Electronics Industries Association standard codes for positioning, straight-cut, and contouring control systems.

ELECTROMAGNETIC INTERFERENCE (EMI) – Unwanted electrical energy or noise induced in the circuits of a device, owing to the presence of electromagnetic fields.

EMULATOR – A device or program which behaves like another system, and produces identical results.

ENCODER – An electromechanical transducer which produces a serial or parallel digital indication of mechanical angle or displacement.

END EFFECTOR (Robots) – The general term used to describe a gripper or other tool used on a robot.

END OF BLOCK CHARACTER – 1. A character indicating the end of a block of tape information. Used to stop the tape reader after a block has been read. 2. The typewriter function of the carriage return when preparing machine control tapes.

END OF PROGRAM – **A miscellaneous function (M02)** indicating completion of a workpiece. (Stops spindle, coolant, and feed after completion of all commands in the block. Used to reset control and/or machine.)

END OF TAPE – **A miscellaneous function (M30)** which stops spindle, coolant and feed after completion of all commands in the block. (Used to reset control and/or machine.)

END POINT – An extremity of a span.

ERROR – The difference between the indicated and desired values of a measured signal.

ERROR DETECTING – A data code in which each acceptable term conforms to certain rules, such that if transmission or processing errors occur, false results can be detected.

ERROR SIGNAL – Difference between the output and input signals in a servo system.

EXCLUSIVE OR – A logical operator, which has the property such that if X and Y are two logic variables, then the function is defined by the following table:

X	Y	Function
0	0	0
0	1	1
1	0	1
1	1	0

The logical operator is usually represented in electrical notation by an encircled plus sign "+". There is no equivalent FORTRAN symbol.

EXECUTE – To carry out an instruction or to run a program.

EXECUTIVE – Software which controls the execution of programs in the computer, based on established priorities and real-time or demand requirements.

EXTENDED ARITHMETIC ELEMENT – A CPU logic element, which provides hardware implemented multiply, divide, and normalise functions.

F

FEEDBACK – The signal or data fed back to a commanding unit from a controlled machine or process to denote its response to the command signal. The signal representing the difference between actual response and desired response that is used by the commanding unit to improve performance of the controlled machine or process.

FEEDBACK CONTROL – Action in which a measured variable is compared to its desired value, with a function of the resulting error signal used as a corrective command.

FEEDBACK DEVICE – An element of a control system which converts linear or rotary motion to an electrical signal for comparison to the input signal, e.g., resolver, encoder, inductosyn.

FEEDBACK LOOP – A closed signal path, in which outputs are compared with desired values to obtain corrective commands.

FEEDBACK RESOLUTION – The smallest increment of dimension that the feedback device can distinguish and reproduce as an electrical output.

FEEDBACK SIGNAL – The measurement signal indicating the value of a directly controlled variable, which is compared with a setpoint to generate a correction command.

FEED ENGAGE POINT – The point where the motion of the Z axis changes from rapid traverse to a programmed feed (usually referred to as the "R" dimension).

FEEDFORWARD (ANTICIPATORY) CONTROL – Action in which information concerning upstream conditions is converted into corrective commands to minimise the effect of the disturbances.

FEED FUNCTION – The relative motion between the tool or instrument and the work due to motion of the programmed axis or axes.

FEEDRATE BY-PASS – A function directing the control system to ignore programmed feedrate and substitute selected operational rate.

FEEDRATE NUMBER – A coded number read from the tape which describes the feedrate function. Usually denoted as the "F" word.

FEEDRATE OVERRIDE – A variable manual control function directing the control system to reduce or increase the programmed feedrate.

FINAL CONTROL ELEMENT – A valve, motor, or other device which directly changes the value of the manipulated variable.

FIRMWARE – Programs or instructions stored in read only memories.

FIRST GENERATION – 1. In the NC industry, the period of technology associated with vacuum tubes and stepping switches. 2. The period of technology in computer design utilising vacuum tubes, electronics, off-line storage on drum or disk, and programming in machine language.

FIXED BLOCK FORMAT – A format in which the number and sequence of **words** and **characters** appearing in successive **blocks** is constant.

FIXED HEADS – Rigidly mounted reading and writing transducers on bulk memory devices.

FIXED SEQUENCE FORMAT – A means of identifying a word by its location in a block of information. Words must be presented in a specific order and all possible words preceding the last desired word must be present in the block.

FLIP FLOP – A bi-stable device. A device capable of assuming two stable states. A bistable device which may assume a given stable state depending upon the pulse history of one or more input points and having one or more output points. The device is capable of storing a bit of information; controlling gates; etc. A toggle.

FLOPPY DISK – A flexible disk used for storing information.

FLOW CHART – A graphical representation of a problem or system in which inter-connected symbols are used to represent operations, data, flow, and equipment.

FLUIDICS – The technique of control that uses only a fluid as the controlling medium. All control is performed without moving elements.

FOREGROUND PROCESSING – Execution of real-time or high priority programs, which can pre-empt the use of computing facilities.

FORMAT – The arrangement of data.

FORMAT CLASSIFICATION – A means, usually in an abbreviated notation, by which the motions, dimensional data, type of control system, number of digits, **auxiliary functions**, etc. for a particular system can be denoted.

FORMAT DETAIL – Describes specifically which words of what length are used by a specific system in the **format classification**.

FORTRAN – Acronym for Formula Translator, an algebraic procedure oriented computer language designed to solve arithmetic and logical programs.

FOURTH GENERATION – In the NC industry, the change in technology of control logic to include computer architecture.

FREQUENCY RESPONSE ANALYSIS – A method of analysing systems based on introducing cyclic inputs and measuring the resulting output at various frequencies.

FREQUENCY RESPONSE CHARACTERISTIC – The amplitude and phase relation between steady-state sinusoidal inputs and the resulting sinusoidal outputs.

FULL DUPLEX – Allows the simultaneous transmission of information in both directions.

FULL PROPORTIONAL SERVO – A system with complete proportionality between output and input.

FULL RANGE FLOATING ZERO – A characteristic of a numerical machine tool control permitting the zero point on an axis to be shifted readily over a specified range. The control retains information on the location of "permanent" zero.

G

G CODE – A word addressed by the letter G and followed by a numerical code defining preparatory functions or cycle types in a numerical control system.

GAIN – The ratio of the magnitude of the output of a system with respect to that of the input (the conditions of operation and measurements must be specified, e.g., voltage, current or power).

GATE – A device which blocks or passes a signal depending on the presence or absence of specified input signals.

GAUGE HEIGHT – A predetermined partial retraction point along the Z axis to which the cutter retreats from time to time to allow safe X–Y table travel.

GENERAL PURPOSE COMPUTER – A computer designed and capable of carrying out a wide range of tasks.

GENERAL PURPOSE PROCESSOR – A computer program which carries out computations on the part program and prepares the author location data for a particular part without reference to machines on which it might be made.

GRAPHICS – The use of a computer to interactively create a drawing displayed on a terminal.

GRAY CODE – A binary code, in which successive values differ in one place only.

GROUP TECHNOLOGY – The grouping of machines and of parts based on similarities in production requirements such that the parts may be produced more efficiently.

H

HALF DUPLEX – Allows the transmission of information one way at a time.

HARD COPY – Any form of computer-produced printed document. Also, sometimes punched cards or paper tape.

HARDWARE – Physical equipment.

HEAD – A device, usually a small electromagnet on a storage medium such as magnetic tape or a magnetic drum, that reads, records, or erases information on that medium. The block assembly and perforating or reading fingers used for punching or reading holes in paper tape.

HOUSEKEEPING – The general organisation of programs stored to ensure efficient system response.

HYSTERESIS – The difference between the response of a system to increasing and decreasing signals.

I

IC – Integrated circuit.

INCREMENTAL DIMENSION – A dimension expressed with respect to the preceding point in a sequence of points.

INCREMENTAL FEED – A manual or automatic input of present motion command for a machine axis.

INCREMENTAL PROGRAMMING – Programming using words indicating incremental dimensions.

INCREMENTAL SYSTEM – Control system in which each coordinate or positional dimension is taken from the last position.

INDEXING – Movement of one axis at a time to a precise point from numeric commands.

INDUCTOSYN SCALE – A precision data element for the accurate measurement and control of angles or linear distances, utilising the inductive coupling between conductors separated by a small air gap.

INHIBIT – To prevent an action or acceptance of data by applying an appropriate signal to the appropriate input.

INITIALISE – To cause a program or hardware circuit to return a program, a system, or a hardware device to an original state or to selected points with a computer program.

INPUT – A dependent variable applied to a control unit or system.

INPUT RESOLUTION – The smallest increment of dimension that can be programmed as input to the system.

INSTABILITY – The state or property of a system where there is an output for which there is not corresponding input.

INSTRUCTION – A statement that specifies an operation and the values or locations of its operands.

INSTRUCTION SET – The list of machine language instructions which a computer can perform.

INTEGRATED CIRCUIT (IC) – A combination of interconnected passive and active circuit elements incorporated on a continuous substrate.

INTEGRATOR – A device which integrates an input signal, usually with respect to time.

INTELLIGENT TERMINAL – A terminal which has its own local processing power.

INTERACTIVE GRAPHICS – Ability to carry out graphics tasks with immediate response from the computer.

INTERFACE – 1. A hardware component or circuit for linking two pieces of electrical equipment having separate functions, e.g., tape reader to data processor or control system to machine. 2. A hardware component or circuit for linking the computer to external I/O device.

INTERFEROMETER – An instrument that uses light interference phenomena for determination of wavelength, spectral fine structure, indices of refraction, and very small linear displacements.

INTERLOCK – To arrange the control of machines or devices so that their operation is interdependent in order to assure their proper coordination.

INTERLOCK BY-PASS – A command to temporarily circumvent a normally provided interlock.

INTERPOLATION – 1. The insertion of intermediate information based on assumed order or computation. 2. A function of a control whereby data points are generated between given coordinate positions to allow simultaneous movement of two or more axes of motion in a defined geometric pattern, e.g., linear, circular and parabolic.

INTERPOLATOR – A device which is part of a numerical control system and performs interpolation.

INTERRUPT – A break in the execution of a sequential program or routine, to permit processing of high priority data.

I/O – (Input/Output) Input or output or both.

ITERATION – A set of repetitive computations, in which the output of each step is the input to the next step.

J

JCL – Job control program

JOB – An amount of work to be completed.

JOG – A control function which provides for the momentary operation of a drive for the purpose of accomplishing a small movement of the driven machine.

K

KEYBOARD – The keys of a teletype-writer which have the capability of transmitting information to a computer but not receiving information.

L

LAG – Delay caused by conditions such as capacitance, inertia, resistance or dead time.

LANGUAGE – A set of representations and rules used to convey information.

LAYOUT – A visual representation of a complete physical entity usually to scale.

LEVEL – 1. Formerly a channel of punched tape. 2. The average amplitude of a variable quantity applying particularly to sound or electronic signals expressed in decibels, volts, amperes, or watts. 3. The degree of subordination in a hierarchy.

LIGHT PEN – A photo sensing device similar to an ordinary fountain pen which is used to instruct CRT displays by means of light sensing optics.

LINEAR INTERPOLATION – A function of a control whereby data points are generated between given coordinate positions to allow simultaneous movement of two or more axes of motion in a linear (straight line) path.

LINE PRINTER – A printing device that can print an entire line of characters all at once.

LINKAGE – A means of communicating information from one routine to another.

LOCKOUT SWITCH – A switch provided with a memory, which protects the contents of designated segments from alteration.

LOG – A detailed record of actions for a period of time.

LOG OFF – The completion of a terminal session.

LOG ON – The beginning of a terminal session.

LOGIC – 1. Electronic devices used to govern a particular sequence of operations in a given system. 2. Interrelation or sequence of facts or events when seen as inevitable or predictable.

LOGIC LEVEL – The voltage magnitude associated with signal pulses representing ONES and ZEROS in binary computation.

LOOP TAPE – A short piece of tape, containing a complete program of operation, with the ends joined.

LSI – Large Scale Integration – A large number of interconnected integrated circuits manufactured simultaneously on a single slice of semi-conductor material.

M

MACHINE LANGUAGE – A language written in a series of bits which are understandable by, and therefore instruct, a computer. The "first level" computer language, as compared to a "second level" assembly language or a "third level" compiler language.

MACHINE PROGRAM – An ordered set of instructions in automatic control language and format recorded on appropriate input media and sufficiently complete to effect the direct operation of an automatic control system.

MACHINING CENTRE – A machine tool, usually numerically controlled, capable of automatically drilling, reaming, tapping, milling and boring multiple faces of a part and often equipped with a system for automatically changing cutting tools.

MACRO – A source language instruction from which many machine language instructions can be generated (see compiler language).

MAGNETIC CORE – An element for switching or storing information on magnetic memory elements for later use by a computer.

MAGNETIC CORE STORAGE – The process of storing information on magnet memory elements for later use by a computer.

MAGNETIC DISK STORAGE – A storage device or system consisting of magnetically coated metal disks.

MAINFRAME – See central processing unit.

MANAGEMENT INFORMATION SERVICE (MIS) – An information feedback system from the machine to management and implemented by a computer.

MANUAL DATA INPUT (MDI) – A means of inserting data manually into the control system.

MANUAL FEEDRATE OVERRIDE – Device enabling operator to reduce or increase the feedrate.

MANUAL PART PROGRAMMING – The manual preparation of a manuscript in machine control language and format to define a sequence of commands for use on an NC machine.

MANUSCRIPT – Form used by a part programmer for listing detailed manual or computer part programming instructions.

MEMORY – A device or media used to store information in a form that can be understood by the computer hardware.

MEMORY, BULK – Any non-programmable large memory, i.e., drum, disk.

MEMORY CYCLE TIME – The minimum time between two successive data accesses from a memory.

MEMORY PROTECT A technique of protecting stored data from alteration, using a guard bit to inhibit the execution of any modification instruction.

MICROPROCESSOR – A single integrated circuit which forms the basic element of a computer.

MICROPROGRAMMING – A programming technique in which multiple instruction operations can be combined for greater speed and more efficient memory use.

MICROSECOND – One millionth of a second.

MILLISECOND – One thousandth of a second.

MISCELLANEOUS FUNCTION – An off–on function of a machine such as Clamp or Coolant on. (See Auxiliary Function).

MNEMONIC – An alphanumeric designation, designed to aid in remembering a memory location or computer operation.

MODEM – A contraction of modulator demodulator. The term may be used with two different meanings: 1. The modulator and the demodulator of a modem are associated at the same end of a circuit. 2. The modulator and the demodulator of a modem are associated at the opposite ends of a circuit to form a channel.

MODULE – An independent unit which may be used on its own or in conjunction with other units to form a complete entity.

MONITOR – A device used for observing or testing the operations of a system.

MOVABLE HEADS – Reading and writing transducers on bulk memory devices which can be positioned over the data locations.

MSI – Medium Scale Integration. (See LSI.) Smaller than LSI, but having at least 12 gates or basic circuits with at least 100 circuit elements.

MULTIPLEXER – A hardware device which handles multiple signals over a single channel.

N

NAND – A combination of the Boolean logic functions NOT and AND.

NAND GATE – A component which implements the NAND function.

NANOSECOND – One thousandth of one microsecond.

NEGATIVE LOGIC – Logic in which the more negative voltage represents the one (1) state; the less negative voltage represents the zero (0) state.

NIXIE LIGHT OR TUBE – A glow lamp which converts a combination of electrical impulses into a visible number.

NOISE – An extraneous signal in an electrical circuit capable of interfering with the desired signal. Loosely, any disturbance tending to interfere with the normal operation of a device or system.

NOR GATE – A component which implements the NOR function.

NOT – A logic operator having property that if P is a logic quantity then quantity "NOT P" assumes values as defined in the following table:

P	NOT P
0	1
1	0

The NOT operator is represented in electrical notation by an overline, e.g., $\bar{P}$ and in FORTRAN by a minus sign "–" in a Boolean expression.

NUMERICAL CONTROL (NC) – A technique of operating machine tools or similar equipment, in which motion is developed in response to numerically coded commands.

NUMERICAL DATA – Data in which information is expressed by a set of numbers that can only assume discrete values.

O

OBJECT PROGRAM – The coded output of an assembler or compiler.

OCTAL – A characteristic of a system in which there are eight elements, such as a numbering system with a radix of eight.

OFF-LINE – Operating software or hardware not under the direct control of a central processor, or operations performed while a computer is not monitoring or controlling processes or equipment.

OFFSET – The steady-state deviation of the controlled variable from a fixed setpoint.

ON-LINE – A condition in which equipment or programs are under direct control of a central processor.

ONE – One of the two symbols normally employed in binary arithmetic and logic, indicating binary one and the true condition, respectively.

OPEN LOOP – A signal path without feedback.

OPEN LOOP SYSTEM – A control system that has no means of comparing the output with the input for control purposes (no feedback).

OPERATING SYSTEM – Software which controls the execution of computer programs and the movement of information between peripheral devices.

OPTIMISATION – A process whose object is to make one or more variables assume, in the best possible manner, the value best suited to the operation in hand, dependent on the values of certain other variables which may be either predetermined or sensed during the operation.

OPTIMISE – To establish control parameters which maximise or minimise the value of performance.

OPTIONAL STOP – **A Miscellaneous Function** command similar to "Program Stop" except that the control ignores the command unless the operator has previously pushed a button to validate the command (M01).

OR – A logic operator having the property that if P and Q are logic quantities then the quantity "P or Q" assumes values as defined by the following table:

P	Q	P OR Q
0	0	0
0	1	1
1	0	1
1	1	1

The OR operator is represented in both electrical and FORTRAN terminology by a "+", i.e., P + Q.

OR GATE – A device which implements the OR function.

ORIENTATION (Robots) – The angular position of the wrist axes.

OUTPUT – Dependent variable signal produced by a transmitter, control unit or other device.

OUTPUT IMPEDANCE – The impedance presented by a device to the load.

OUTPUT SIGNAL – A signal delivered by a device, element, or system.

OVERLAY – A technique of repeatedly using the same area of computer store when actioning different stages of a problem.

OVERSHOOT – The amount that a controlled variable exceeds its desired value after a change of input.

P

PARABOLA – A plane curve generated by a point moving so that its distance from a fixed second point is equal to its distance from a fixed line.

PARABOLIC INTERPOLATION – Control of cutter path by interpolation between three (3) fixed points by assuming the intermediate points are on a parabola.

PARALLEL – The simultaneous transfer and processing of all bits in a unit of information.

PARAMETER – A characteristic of a system or device, the value of which serves to distinguish various specific states.

PARITY CHECK – A test of whether the number of ONES or ZEROS in an array of binary digits is odd or even to detect errors in a group of bits.

PART PROGRAM – An ordered set of instructions in a language and in a format required to cause operations to be effected under automatic control, which is either written in the form of a machine program on an input media or prepared as input data for processing in a computer to obtain a machine program.

PART PROGRAMMER – A person who prepares the planned sequence of events for the operation of a numerically controlled machine tool.

PASSWORD – A word the operator must supply in order to meet the security requirements and gain access to the computer.

PATCH – Temporary coding used to correct or alter a routine, or a term used in CAD.

PERIPHERAL – Auxiliary equipment used for entering data into or receiving data from a computer.

PERIPHERAL EQUIPMENT – The auxiliary machines and storage devices which may be placed under control of the central computer and may be used on-line or off-line, e.g., card reader and punches, magnetic tape feeds, high speed printers, CRTs and magnetic drums or disks.

PICOSECOND – One millionth of one microsecond.

PITCH (Robots) – A rotation of the payload or tool about a horizontal axis on the end of a robot arm which is perpendicular to the longitudinal axis of the arm.

PLANNING SHEET – A list of operations for the manufacture of a part, prepared before the part program.

PLOTTER – A device used to make a drawing of a display.

POINT-TO-POINT CONTROL SYSTEM – An NC system which controls motion only to reach a given end point but exercises no path control during the transition from one end point to the next.

POLAR AXES – The fixed lines from which the angles made by radius vectors are measured in a polar coordinates system.

POLAR COORDINATES – A mathematical system for locating a point in a plane by the length of its radius vector and the angle this vector makes with a fixed line.

POSITION READOUT – A display of absolute slide position as derived from a position feedback device (transducer usually) normally attached to the lead screw of the machine. (See Command Readout.)

POSITION SENSOR – A device for measuring a position, and converting this measurement into a form convenient for transmission.

POSITION STORAGE – The storage media in an NC system containing the coordinate positions read from tape.

POSITIVE LOGIC – Logic in which the more positive voltage represents the one (1) state.

POST-PROCESSOR – A computer program which adapts the output of a processor into a machine program for the production of a part on a particular combination of machine tool and controller.

PRECISION – The degree of discrimination with which a quantity is stated, e.g., a three-digit numeral discriminates among 1000 possibilities. Precision is contrasted with accuracy, i.e., a quantity expressed with 10 decimal digits of precision may only have one digit of accuracy.

PREPARATORY FUNCTION – An NC command on the input tape changing the mode of operation of the control. (Generally noted at the beginning of a block by "G" plus two digits.)

PREPROCESSOR – A computer program which prepares information for processing.

PREVENTATIVE MAINTENANCE – Maintenance specifically designed to identify potential faults before they occur.

PRINTED CIRCUIT – A circuit for electronic components made by depositing conductive material in continuous paths from terminal to terminal on an insulating surface.

PROCESSOR – A computer program which processes information.

PROGRAM – A plan for the solution of a problem. A complete program includes plans for the transcription of data, coding for the computer, and plans for the absorption of the results into the system. The list of coded instructions is called a routine. To plan a computation or process from the asking of a question to the delivery of the results, including the integration of the operation into an existing system. Thus, programming consists of planning and coding, including numerical analysis, systems analysis, specification of printing formats, and any other functions necessary to the integration of a computer in a system.

PROGRAMMABLE – Capable of being set to operate in a specified manner, or of accepting remote setpoint or other commands.

PROGRAMMED ACCELERATION – A controlled velocity increase to the programmed feedrate of an NC machine.

PROGRAMMED DWELL – The capability of commanding delays in program execution for a programmable length of time.

PROGRAM STOP – A **Miscellaneous Function** (M00) command to stop the spindle, coolant and feed after completion of the dimensional move commanded in the **block**. To continue with the remainder of the program, the operator must initiate a restart.

PROTOCOL – Set of rules governing message exchange between two devices.

PUNCHED PAPER TAPE – A strip of paper on which characters are represented by combinations of holes.

PULSE – A short duration change in the level of a variable.

Q

QUADRANT – Any of the four parts into which a plane is divided by rectangular coordinate axes lying in that plane.

QUADRATURE – Displaced 90 degrees in phase angle.

R

R DIMENSION – (See Feed Engage Point).

RANDOM ACCESS MEMORY (RAM)   A storage unit in which direct access is provided to information, independent of memory location.

RASTER DISPLAY – A display in which the entire display surface is scanned at a constant refresh rate.

RASTER SCAN – Line-by-line sweep across the entire display surface to generate elements of a display image.

READ – To acquire data from a source. To copy, usually from one form of storage to another, particularly from external or secondary storage to internal storage. To sense the meaning of arrangements of hardware. To sense the presence of information on a recording medium.

READER – A device capable of sensing information stored in off-line memory media (cards, paper tape, magnetic tape) and generating equivalent information in an on-line memory device (register, memory locations).

READ ONLY MEMORY (ROM) – A storage device generally used for control program, whose content is not alterable by normal operating procedures.

REAL TIME CLOCK – The circuitry which maintains time for use in program execution and event initiation.

REAL TIME OPERATION – Computer monitoring, control, or processing functions performed at a rate compatible with the operation of physical equipment or processes.

REFERENCE BLOCK – A block within an NC program identified by an "O" or "H" in place of the word address "N" and containing sufficient data to enable resumption of the program following an interruption. (This block should be located at a convenient point in the program which enables the operator to reset and resume operation.)

REFRESH – CRT display technology which requires continuous restroking of the display image.

RELOCATABLE POINT/SEQUENCE OF POINT (Robots) – A point or sequence in a robot which can be relocated in space.

REPAINT – Redraws a display on a CRT to reflect its current status.

REPEATABILITY – The closeness of agreement among multiple measurements of an output, for the same value of the measured signal under the same operating conditions, approaching from the same direction, for full range traverses.

REPRODUCIBILITY – The closeness of agreement among repeated measurements of the output for the same value of input, made under the same operating conditions over a period of time, approaching from either direction.

RESOLUTION – 1. The smallest distinguishable increment into which a signal or picture, etc. is divided in a device or system. 2. The minimum positioning motion which can be specified.

RESOLVER – 1. A mechanical to electrical transducer (see Transducer) whose input is a vector quantity and whose outputs are components of the vector. 2. A transformer whose coupling may be varied by rotating one set of windings relative to another. It consists of a stator and rotor, each having two distributed windings 90 electrical degrees apart.

RETROFIT – Work done to an existing machine tool from simply adding special jigs or fixtures to the complete re-engineering and manufacturing, and often involving the addition of a numerical control system.

ROBOT – An automatic device which performs functions ordinarily ascribed to human beings.

ROLL (Robots) – A rotation of the payload or tool about the longitudinal axis of the wrist.

ROUTINE – A series of computer instructions which performs a specified task.

RUN – The execution of a program on a computer.

S

SAMPLE AND HOLD – A circuit used to increase the interval during which a sampled signal is available, by maintaining an output equal to the most recent input sample.

SAMPLES DATA – Data in which the information content can be, or is, ascertained only at discrete intervals of time. (Can be analog or digital.)

SAMPLING PERIOD – The interval between observations in a periodic sampling control system.

SCALE – To change a quantity by a given factor, to bring its range within prescribed limits.

SCALE FACTOR – A coefficient used to multiply or divide quantities in order to convert them to a given magnitude.

SCHEDULE – A programme or timetable of planned events or of work.

SECOND GENERATION – 1. In the NC industry, the period of technology associated with transistors (solid state). 2. The period of technology in computer design utilising solid-state circuits, off-line storage, and significant development in software, the assembler.

SECURITY – Prevention of unauthorised access to information or programs.

SENSITIVITY – The ratio of a change in steady state output to the corresponding change of input, often measured in percentage of span.

SENSOR – A unit which is actuated by a physical quantity and which gives a signal representing the value of that physical quantity.

SEQUENCE (Robots) – Part of a robot program which consists of a point or series of points the performance of which will be dependent on defined input/flag conditions existing.

SEQUENCE CONTROL – A system of control in which a series of machine movements occurs in a devised order, the completion of one movement initiating the next, and in which the extent of the movements is not specified by numeric data.

SERIAL – The transfer and processing of each bit in a unit of information, one at a time.

SERVO AMPLIFIER – The part of the servo system which increases the error signal and provides the power to drive the machine slides or the servo valve controlling a hydraulic drive.

SETPOINT – The position established by an operator as the starting point for the program on an NC machine.

SIGN – The symbol or bit which distinguishes positive from negative numbers.

SIGNAL – Information conveyed between points in a transmission or control system, usually as a continuous variable.

SIGNIFICANT DIGIT – A digit that contributes to the precision of a numeral. The number of significant digits is counted beginning with the digit contributing the most value, called the most significant digit, and ending with the one contributing the least value, called the least significant digit.

SIMULATOR – A device or computer program that performs simulation.

SKEWING – Refers to time delay or offset between any two signals in relation to each other.

SOFTWARE – The collection of programs, routines, and documents associated with a computer.

SOURCE IMPEDANCE – The impedance presented to the input of a device by the source.

SOURCE LANGUAGE – The symbolic language comprising statements and formulas used to specify computer processing. It is translated into object language by an assembler or compiler, and is more powerful than an assembly language in that it translates one statement into many items (see macro).

STABILITY – Freedom from undesirable deviation, used as a measure of process controllability.

STANDBY POWER SUPPLY – An energy generation or storage system that can permit equipment to operate temporarily or shut down in an orderly manner.

STATIC GAIN – The ratio of steady-state output to input change.

STEADY STATE – A characteristic or condition exhibiting only negligible change over an arbitrarily long period of time.

STEPPING MOTOR – A bi-directional permanent magnet motor which turns in finite steps.

STEP RESPONSE – The time response of an instrument subjected to an instantaneous change in input.

STEP RESPONSE TIME – The time required for an element output to change from an initial value to a specified percentage of a steady state, either before or in the absence of overshoot, after an input step change.

STORAGE – A memory device in which data can be entered and held, and from which it can be retrieved.

STORAGE TUBE – A CRT which retains an image for a considerable period of time without redrawing.

STRAIGHT CUT SYSTEM – A system which has feedrate control only along the axes and can control cutting action only along a path parallel to the linear (or circular) machine ways.

SUB PROGRAM – A segment of a machine program which can be called into effect by the appropriate machine control command.

SUBROUTINE – A series of computer instructions to perform a specific task for many other routines. It is distinguishable from a main routine in that it requires, as one of its parameters, a location specifying where to return to the main program after its function has been accomplished.

SUMMING POINT – A point at which signals are added algebraically.

SYNCHRO – A transformer having a polyphase primary winding and single phase secondary winding which can be rotated. The voltage induced into the secondary may be controlled in phase by turning the secondary coil.

SYNCHRONOUS – A fixed rate transmission of information synchronised by a clock for both receiver and sender.

SYNTAX – The rules which govern the structure of words and expressions in a language.

T

TABLET – An input device which allows digitised coordinates to be indicated by stylus position.

TACHOMETER – A speed measuring instrument generally used to determine revolutions per minute. In NC it is used as a velocity feedback device.

TAPE – A magnetic or perforated paper medium for storing information.

TAPE LEADER – The front or lead portion of a tape.

TAPE PREPARATION – The act of translating command information into punched or magnetic tape.

TAPE TRAILER – The trailing end portion of a tape.

TASK – A unit of work.

TEACH (Robots) – The mode by which a robot is driven to required points in space for programming.

TERMINAL – A device by which information may be entered or extracted from a system or communication network.

THIRD GENERATION – 1. In the NC industry, the period of technology associated with integrated circuits. 2. The period of technology in computer design utilising integrated circuits, core memory, advanced subroutines, time sharing, and fast core access.

THRESHOLD – The minimum value of a signal required for detection.

TIME CONSTANT – For a first order system, the time required for the output to complete 63.2% of the total rise or decay as a result of a step change of the input.

TIME SHARING – The interleaved use of a sequential device, to provide apparently simultaneous service to a number of users.

TOGGLE – A flip-flop or two-position switch.

TOOL CENTRE POINT (Robots) – The real or imaginary offset point defined in relation to the tool mounting plate of a robot which moves in a straight line between programmed points and at the programmed velocity in controlled path machines.

TOOL FUNCTION – A tape command identifying a tool and calling for its selection. The address is normally a "T" word.

TOOL LENGTH COMPENSATION – A manual input means which eliminates the need for preset tooling and allows programmer to program all tools as if they are of equal length.

TOOL OFFSET – 1. A correction for tool position parallel to a controlled axis. 2. The ability to reset tool position manually to compensate for tool wear, finish cuts and tool exchange.

TOOLPATH – The geometry of the path a tool will follow to machine a component.

TOOLPATH FEEDRATE – The velocity, relative to the workpace, of the tool reference point along the author path, usually expressed in units of length per minute or per revolution.

TRACK – The portion of a moving storage medium, such as the drum, tape or disc, that is accessible to a given reading head position.

TRANSFER FUNCTION – An expression relating the output of a linear system to the input.

TRUNCATE – To terminate a computational process in accordance with some rule, e.g., to end the evaluation of a power series at a specified term.

TRUTH TABLE – A matrix that describes a logic function by listing all possible combinations of inputs, and indicating the outputs for each combination.

TUNING – The adjustment of coefficients governing the various modes of control.

TURNING CENTRE – A lathe type numerically controlled machine tool capable of automatically boring, turning outer and inner diameters, threading, facing multiple diameters and faces of a part and often equipped with a system for automatically changing or indexing cutting tools.

TURN KEY SYSTEM – A term applied to an agreement whereby a supplier will install an NC or computer system so that he has total responsibility for building, installing, and testing the system.

V

VARIABLE (Robots) – An ability to count events.

VARIABLE BLOCK FORMAT – Tape format which allows the number of words in successive blocks to vary.

VECTOR – A quantity that has magnitude, direction and sense and that is commonly represented by a directed line segment whose length represents the magnitude and whose orientation in space represents the direction.

VECTOR FEEDRATE – The resultant feedrate which a cutter or tool moves with respect to the work surface. The individual slides may move slower or faster than the programmed rate; but the resultant movement is equal to the programmed rate.

VOLATILE STORAGE – A memory in which data can only be retained while power is being applied.

W

WINDUP – Lost motion in a mechanical system which is proportional to the force or torque applied.

WIRE-FRAME – A 3-dimensional drawing created by the projection of the points of intersection of the geometry.

WORD ADDRESS FORMAT – Addressing each word in a block by one or more characters which identify the meaning of the word.

WORD LENGTH – The number of bits or characters in a word.

WORLD COORDINATES (Robots) – The coordinate system by which a point in space is defined in three cartesian coordinates and three orientation or polar coordinates.

WRIST (Robots) – The element of a robot which applies orientation to a tool.

X

X AXIS – Axis of motion that is always horizontal and parallel to the work-holding surface.

Y

Y AXIS – Axis of motion that is perpendicular to both the X and Z axes.

YAW (Robots) – A rotation of a payload or tool about a vertical axis that is perpendicular to the pitch axis of the wrist.

Z

Z AXIS – Axis of motion that is always parallel to the principal spindle of the machine.

ZERO – One of the two symbols normally employed in binary arithmetic and logic, indicating the value zero and the false condition, respectively.

ZERO OFFSET – A characteristic of a numerical machine tool control permitting the zero point on an axis to be shifted readily over a specified range. (The control retains information on the location of the "permanent" zero.)

ZERO SHIFT – A characteristic of a numerical machine tool control permitting the zero point on an axis to be shifted readily over a specified range. (The control does **not** retain information on the location of the "permanent" zero.)

ZERO SUPPRESSION – The elimination of non-significant zeros to the left of significant digits usually before printing.

ZERO SYNCHRONISATION – A technique which permits automatic recovery of a precise position after the machine axis has been approximately positioned by manual control.

[COURTESY OF THE NUMERICAL ENGINEERING SOCIETY (UK)]

Selected Bibliography

Books

I. Aleksander, Designing intelligent systems – an introduction. Kogan Page, 1984. ISBN 0-85038-860-0.

ASM International, High speed machining – solutions for productivity. 1990. ISBN 0-87170-367-X.

B.G. Batchelor, D.A. Hill, D.C. Hodgson, Automated visual inspection. IFS, 1985. ISBN 0-903608-68-51.

D. Bennett, Production systems design. Butterworth, 1986. ISBN 0-408-01546-2.

C.W. Besant, C.W. Lui, Computer-aided design and manufacture. Ellis Horwood, 1986. ISBN 0-85312-909-6.

J. Cullen, J. Hollingum, Implementing total quality. IFS/Springer-Verlag, 1987. ISBN 0-948507-65-9.

D. Gibbs, An introduction to CNC machining. Cassell, 1987. ISBN 0-304-31412-9.

D. Gibbs, CNC part programming: a practical guide. Cassell, 1987. ISBN 0-304-31355-6.

D.L. Goetsch, Advanced manufacturing technology. Delmar, 1990. ISBN 0-8273-3786-8.

M.J. Haigh, An introduction to computer-aided design and manufacture. Blackwell, 1985. ISBN 0-632-01242-0.

J. Hartley, FMS at work. IFS, 1984. ISBN 0-903608-62-6.

J. Howlett, Tools of total quality. Chapman & Hall, 1991. ISBN 0-412-37690-3.

H.B. Kief, Flexible automation. Becker, 1986. ISBN 0-9512010-0-X.

A. Kochan, D. Cowan, Implementing CIM. IFS/Springer-Verlag, 1986. ISBN 0-948507-20-9.

B. Leatham-Jones, Introduction to computer numerical control. Pitman, 1986. ISBN 0-273-02402-7.

D.K. Macbeth, Advanced manufacturing – strategy and management. IFS/Springer-Verlag, 1989. ISBN 1-85423-039-5.

A.J. Medland, P. Burnett, CADCAM in practice. Kogan Page, 1986. ISBN 0-85038-817-1.

A.J. Medland, G. Mullineux, Principles of CAD – a coursebook. Kogan Page, 1988. ISBN 1-85091-534-2.

J.S. Oakland, Statistical process control – a practical guide. Heinemann, 1986. ISBN 0-434-91475-4.

D. Parrish, Flexible manufacturing. Butterworth–Heinemann, 1990. ISBN 0-7506-10115.

J. Pusztai, M. Sava, Computer numerical control. Reston, 1983. ISBN 0-8359-0924-7.

P.G. Ranky, The design and operation of FMS. IFS, 1983. ISBN 0-903608-44-8.

W.S. Robertson, Lubrication in practice. Macmillan, 1987. ISBN 0-333-34978-4.

W.S. Seames, Computer numerical control: concepts and programming. Delmar, 1990. ISBN 0-8273-3782-5.

G.T. Smith, Advanced machining – the handbook of cutting technology. IFS/Springer-Verlag, 1989. ISBN 1-085423-022-0.

K.J. Stout, Quality control in automation. Kogan Page, 1985. ISBN 0-85038-936-4.

G.E. Thyer, Computer numerical control of machine tools. Heinemann, 1988. ISBN 0-434-91959-4.

J. Woodwark, Computing shape. Butterworths, 1986. ISBN 0-408-01402-4.

Magazines and Journals

American Machinist (monthly)
Industrial Engineering (monthly) (Marcel Dekker Inc)
Industrial Robot (quarterly) (IFS)
Integrated Manufacturing Systems (quarterly)
International Journal of Advanced Manufacturing Technology (quarterly) (Springer International)
International Journal of Machine Tools and Manufacture (quarterly) (Pergamon Press)
International Journal of Production Research (monthly) (Taylor Francis)
Journal of Engineering for Industry (quarterly)
Journal of Engineering Manufacture (quarterly)
Logistics World (quarterly) (IFS)
Machinery and Production Enginerring (monthly)
Networking Production (monthly)
Tribology International (bi-monthly) (Butterworth–Heinemann)
Wear (seven/annum) (Elsevier Sequoia)

Company Addresses

UK Head Office   *International Head Office*

William Asquith (1981) Ltd
Highroad Well Works
Gibbet Street
Halifax
W. Yorks
HX2 OAP
Tel: 0422 367771

Boko Machine Tools (UK) Ltd
Kingfield Industrial Estate
Kingfield Road
Coventry
West Midlands
CV1 4DW
Tel: 0203 228447/8

Bohner & Kohle
GmbH & Co
Maschinenfabrik
Weilstrasse 4–10
D-7300 Esslingen/Neckar
Postfach 67
Germany
Tel: 0711-3901-0

Bostomatic UK Ltd
7 Prospect Way
Butler's Leap
Rugby
CV21 3UU
Tel: 0788 73865

Boston Digital Corp
Granite Park
Milford
MA 01757
USA
Tel: (508) 473–4561

Bridgeport Machines Ltd
PO Box 22
Forest Road
Leicester
LE5 OFJ
Tel: 0533 531122

Bridgeport Machines Inc
500 Lindley Street
Bridgeport
CT 06606
USA

Butler Newall Ltd
(Electronic Accuracy Systems)
Westholme Road
Halifax
HX1 4JR
Tel: 0422 331144

UK Head Office   *International Head Office*

BYG Systems Ltd
Highfields Science Park
University Boulevard
Nottingham
NG7 2QP
Tel: 0602 252221

Carne (UFM) Ltd   **Ibag Zürich AG**
Swan Works   Glattalstrasse 138
416–418 London Road   CH-8052 Zürich
Isleworth   Switzerland
Middlesex   Tel: 01 301 0020
TW7 5AE
Tel: 081 5601182

Centreline Machine Tool Co. Ltd
Newton Road
Hinckley
Leics
LE10 3DS
Tel: 0455 618012

Cimcool Division Head Office   **Cimcool Industrial Products**
Cincinnati Milacron Ltd   PO Box 463
Maybrook Road   3130 Al-Vlaardingen
Castle Vale Industrial Estate   Netherlands
Sutton Coldfield   460 0660
West Midlands
B76 8BB
Tel: 021 351 1891

The Cimulation Centre
Avon House
PO Box 46
Chippenham
Wiltshire
SN15 1JH
Tel: 0249 650316

Cincinnati Milacron UK Ltd   **Cincinnati Head Quarters**
PO Box No. 505   Oakley Complex
Kingsbury Road   4701 Marburg Avenue
Birmingham   Cincinnati
B24 0QU   Ohio 45209
Tel: 021 351 3821   USA
                                                           Tel: 513 841 8100

Cranfield Precision Engineering Ltd (CUPE)
Cranfield Institute of Technology
Cranfield
Bedford
MK43 0AL
Tel: 0234 752789

Crawford Collets Ltd
Tower Hill Works
Witney
Oxfordshire
OX8 5DS
Tel: 0993 703931

UK Head Office

International Head Office

Dean Smith & Grace Ltd
Keighley
West Yorkshire
BD1 4PG
Tel: 0535 605261

Monarch M/C Tool Co
615 North Oak Street
PO Box 668
Sidney
Ohio 45365
USA
Tel: 513 492 4111

DeBeers Industrial Diamond Division Ltd
Charters
Sunninghill
Ascot
Berks
SL5 9PX
Tel: 0990 23456

Devlieg Microbore Tooling Co
Leicester Road
Lutterworth
Leics
LE17 4HE
Tel: 0455 553030

Eclipse Magnetics Ltd
Vulcan Road
Sheffield
S9 1EW
Tel: 0742 560600

Federal-Mogul
Westwind Air Bearings Ltd
Trading Park
Holton Heath
Poole
Dorset
BH16 6LN
Tel: 0202 622565

Ferranti Industrial Electronics Ltd
Dunsinane Avenue
Dundee
DD2 3PN
Tel: 0382 89311

Flexible Manufacturing Technology Ltd (FMT)
Carden Avenue
Hollingbury
Brighton
BN1 8AU
Tel: 0273 507255

GE Fanuc Automation
1 Fairways
Pitfield
Kiln Farm
Milton Keynes
MK11 3EE
Tel: 0908 260200

GE Fanuc Automation Europe SA
Zone Industrielle Echternach
Grand-Duche de Luxembourg
Tel: 728372

UK Head Office

International Head Office

Gildemeister (UK) Ltd
Unitool House
Camford Way
Sundon Park
Luton
LU3 3AN
Tel: 0582 570661

Gildemeister Aktiengesellschaft
Morsestrasse 1
D-4800 Bielefeld 11
Germany
Tel: 05205 7510

Hahn & Kolb (GB) Ltd
6 Forum Drive
Leicester Road
Rugby
Warks
CV21 1NY
Tel: 0788 577288

Hahn & Kolb GmbH & Co
Bei Angabe Des
Postfach 106018
W-7000
Stuttgart 10
Germany
Tel: 49 711 945-0

Iscar Tools Ltd
Woodgate Business Park
156 Clapgate Lane
Bartley Green
Birmingham
B32 3DE
Tel: 021 422 4070

Iscar Limited
Box 11
Tefen 24959
Israel
Tel: 0109724970311

AT & T Istel
Industrial Systems Ltd
Highfield House
Headless Cross Drive
Redditch
Worcs
B97 5EG
Tel: 0527 550330

Jones & Shipman PLC
Narborough Road South
PO Box 89
Leicester
LE3 2LF
Tel: 0533 896222

Kennametal Europe
PO Box 29
Kingswinford
West Midlands
DY6 7NP
Tel: 0384 408010

Kennametal Inc International
PO Box 231
Latrobe
PA 15650
USA
Tel: 412 539 4700

Krupp Widia UK Ltd
5 The Valley Centre
Goron Road
High Wycombe
Bucks
Tel: 0494 451845

Krupp Widia GmbH
Münchener Strasse 90
D-4300 Essen 1
Germany
Tel: 0201 725 0

KT-Swasey Ltd
Stafford Park 2
Telford
Shropshire
TF3 3BO
Tel: 0952 290200

KT-Swasey
11000 W. Theodore Trecker Way
Milwaukee
Wisconsin 53214-0277
USA
Tel: 414 476 8300

UK Head Office

International Head Office

LK Tool Company Ltd
East Midlands Airport
Castle Donnington
Derby
DE7 2SA
Tel: 0332 811349

McDonnell Douglas
Information Systems Ltd
Boundary Way
Hemel Hempstead
Herts
HP2 7HU
Tel: 0442 232424

McDonnell Douglas
Information Systems International
18881 Von Karman
Suite 1800
Irvine
California 92715
USA
Tel: 714 724 5600

Micro Aided Engineering Ltd
Bolton Business Centre
44 Lower Bridgeman Street
Bolton
Greater Manchester
BL2 1DG
Tel: 0204 396500

NC Engineering Ltd
1 Park Avenue,
Bushey
Watford
Herts
WD2 2DG
Tel: 0923 243962

NCMT Ltd (Makino)
Ferry Works
Thames Ditton
Surrey
KT7 0QQ
Tel: 081 398 3402

P-E Information Systems Ltd (Hocus)
Park House
Egham
Surrey
TW20 0HW
Tel: 0784 34411

Rank Taylor Hobson Ltd
PO Box 36
New Star Road
Leicester
LE4 7JQ
Tel: 0533 763771

Renishaw Metrology Plc
New Mills
Wotton-under-Edge
Glouce
GL12 8JR
Tel: 0453 844211

UK Head Office

International Head Office

Renishaw Transducer Systems Ltd
Old Town
Wotton-under-Edge
Gloucs
GL12 7DH
Tel: 0453 844302

Ringspan (UK) Ltd
3 Napier Road
Bedford
MK41 0QS
Tel: 0234 42511

Ringspan GmbH
Schaberweg 30–34
6380 Bad Homburg
Germany
Tel: 06172 275-0

Röhm (GB) Ltd
The Albany Boat House
Lower Ham Road
Kingston-upon-Thames
Surrey
KT2 5BB
Tel: 081 549 6647

Sandvik Coromant UK
Manor Way
Halesowen
West Midlands
B62 8QZ
Tel: 021 550 4700

Sandvik AB
81181 Sandviken
Sweden
Tel: 026 260000

Scharmann Machine Ltd
Cannon House
2255 Coventry Road
Sheldon
Birmingham
B26 3NX
Tel: 021 742 4216

Dörries Scharmann GmbH
D-4050 Mönchengladbach 2
Hugo-Junkers Strasses 12–32
Germany
Tel: 02166 454-0

Seco Tools (UK) Ltd
Kinwarton Farm Road
Alcester
Warks
B49 6EL
Tel: 0789 764341

Seco Tools AB
77301 Fabest
Sweden
Tel: 022-340000

Siemens PLC
Sir William Siemens House
Princess Road
Manchester
M20 8UR
Tel: 061 446 5740

Siemens Aktiengesellschaft
Power Engineering & Automation Group
Numerical Controls & Drives For Machine Tools
Division
PO Box 4848
8500 Nuemberg 1
Germany

SMG Ltd
Industrial House
River Dee Business Park
River Lane
Saltney
Chester
CH4 8QY
Tel: 0244 681206

SMT Machine Company AB
PO Box 800
S-721 22 Västeras
Sweden
Tel: +46 (0)21-805120

UK Head Office

International Head Office

Stellram Ltd
Hercules Way
Bowerhill Industrial Estate
Melksham
Wiltshire
SN12 6TS
Tel: 0225 706882

Stellram SA
1260 Nyon
Switzerland
Tel: 022-613101

System 3R (UK) Ltd
2 Duke Street
Princes Risborough
Bucks
HP17 0AT
Tel: 08444 4339

System 3R International AB
Sorterargatan
S-162 26 Vallingby
Sweden
Tel: 46(8)6202000

Tecnomagnetica UK Ltd
18 Riverside Estate
Sir Thomas Longley Road
Frindsbury
Rochester
Kent
ME2 4DP
Tel: 0634 715802

Tecnomagnete SPA
Via Dei Cignoli 9
20151
Milano
Italy

Thame Engineering Co. Ltd
Field End
Thame Road
Long Crendon
Aylesbury
Bucks
HP18 9EJ
Tel: 0844 208050

Walter Cutters & Grinders Ltd
Walker Road
North Moons Industrial Estate
Redditch
Worcs
B98 9HE
Tel: 0527 60281

Walter AG
Postfach 2049
W-7400
Tuzbingen
Germany
Tel: 497071 7010

WDS Wharton LTD
Marlco Works
Hagden Lane
Watford
WD1 8NA
Tel: 0923 226606

Wix & Royd Ltd
77–81 Brighton Road
Redhill
Surrey
RH1 6PS
Tel: 0727 768823

Yamazaki Mazak UK Ltd
Badgeworth Drive
Worcester
WR4 9NF
Tel: 0905 755755

Index

Absolute dimensional positions 12
Acramatic 850 turning centre 4
Acramatic 950 controller 5
Alternating flank infeed technique 33–34
Angular rotation 15
APT (Automatically Programmed Tool) techniques 97, 99
Artificial intelligence (AI) 3
Auxiliary function 40–1
Axis motions
  with machining 20
  without machining 20

Background/parallel programming 2
Block elements 10–11
Block end character 8
Block format 8–10
Block linking 51
Block processing time 80
Block search 10
Blueprint programming 49–51
BS3800 101
Bursting pressures 82

CAD/CAM systems 6, 19, 89–100
  additional hardware items 94
  application factors 93
  archiving techniques 94
  back-up files 91
  back-up techniques 94
  bench-mark test 89
  choice of computerware 93–5
  choice of supplier 93
  computer considerations 89
  data storage 92
  document management 95
  hard disk capacity 91
  hardware requirements 91
  implementation program 94
  multi-surface capability 97
  peripheral devices 92
  printers/plotters 95

resolution capacities 95
screen graphics 92
sculptured surfaces 95–100
software applications 92
software considerations 89
up-grading 92
window technology 95
Canned cycles 70–6
Check surfaces 100
Circular interpolation 21
CNC controllers 3–5
CNC machine tools, determination of accuracy and repeatability of positioning 101
CNC programming
  fundamentals of 6, 8–76
  methods of 2
  sequencing 5–6
Company addresses 127
Compensation values, changing 64–6
Computer-aided design and manufacure. See CAD/CAM
Computer-aided draughting and design (CADD) 93
Computer-assisted part programming (CAPP) 2
Computer memory 89
Constant cutting speed 31
Constant-depth cutting 97–100
Contour errors 67–8
Contouring cycle programming 49–50
Conversational language program (CAP) 1
Conversational programming 2, 49–51
Corner milling, effect of servo-lag and gain 78–9
Curve design 95
Curve fitting 95
Cutter design, and stiffness 76–7
Cutter radius compensation (CRC) 59, 62–70
  special case problems 68–70
Cutter transformations 15–20
Cylindrical interpolation 27

Datum shifting 17–18
Default condition 10
Deletable blocks 11

Digitising 2, 82–9
  CNC system performance 85
  efficiency considerations 87
  moulds 87–9
  performances 86
  principles of 82–3
  system performance 83–9
  techniques for milling dies and moulds 89
Dimensioning systems 12–14
Direction of compensation, changing 64

Electropilot controller 4
Extended address 11

Feed motions 30–1
Flank infeed technique 33

Gain
  effect on corner milling 78–9
  effect whilst generating circular paths 79–80
Glossary of terms 102–24
Gouge avoidance 97, 100
Group technology (GT) 2, 19

Helical interpolation 25–7
High-speed milling operations, fundamentals of
    76–81
High-speed turning operations 81–2

Incremental dimensional positions 12
Internal semi-circle, milling of 48–9
Interpolation parameters 23
Intersection cutter radius compensation 57–9

Jump functions 10

Large scale integration (LSI) electronic component
    technology 3
Linear interpolation 20–1
Linking of blocks 51
Local/Wide Area Networks (LAN/WAN) 1

Machine tool accuracy and rigidity 76
Machine tool scale system 86
Manual data input (MDI) 41
Manufacturing automation protocol (MAP) 1
Milling operations
  five-axes 99
  high-speed 76–81
  internal semi-circle 48–9
  rectangles using parametric programming 48
  three-axes 97–100
Mirror-imaging 15–17
Miscellaneous functions 38–41
Model stylus considerations 86
Modified flank infeed technique 33
Modularity of hard/software 1–2
Motion blocks 20–30
Motorola 68020 microprocessors 3
Moulds, digitising 87–9
Multiple threads 37–8

NMG (Numerical Master Geometry) 99
Non-Uniform Rational B-Splines (NURBS) 96
NURBS 96

Off-line programming 2
Offset number, changing 64

Parameter calculations 46–7
Parameter definition 44–6
Parameter string 47–8
Parametric programming 2, 44–9
  milling of rectangles using 48
Part programming
  CRC/TNRC in 62
  using "canned cycles" 70–6
Pitch circle diameters (PCDs) 28
Plunge cutting 97–100
Polar coordinates 28–30
Polar dimensional positions 12
Positional measuring system 86
Preparatory functions 20, 24, 31
Printers/plotters 95
Processing speed 80–1
Programmable logic controllers (PLCs) 1
Proportional servo-systems 77

Radial infeed technique 32
Radius programming 24–5
Random access memory (RAM) 89
Reference points 11–2
Reverse cutting 97–100
Reverse engineering 2, 82–9

Scaling 18–20
Sculptured surfaces 95–100
Servo-lag
  effect on corner milling 78–9
  effect whilst generating circular paths 79–80
  problems of 77–80
Set-up support package 5
Shop-floor programming 49–51
Siemens Sinumerik controller 4
Spindle function 40
Splines 96
Standards 101
Stylus performance factors 83–5
Subroutine 41–4
  nesting 43

Thread cutting 31–41
Thread infeed techniques 31–4
Thread on tapered bar 36
Thread programming 34–8
Thread with constant lead 34–6
Tilt angles 99
Tool length compensation 53, 56–7
Tool nose centre 53
Tool nose compensation 53–6
Tool nose radius compensation (TNRC) 56, 62–70
  special case problems 68–70

Tool number 41
Tool offset number 53
Tool offsets
  and their compensations 53-9
  with tool nose radius compensation 56
  without using tool nose compensation 53-6
Tooling, classification 53
Touch-trigger probes 83-5
Touch-trigger probing, digitising of 86
Transversal thread 36
Trickle-feeding 87
Turning operations, high-speed 81-2

User-macros 74-6

Whipping tendency of workpieces 82
Word address format 1, 11
Word address programming 2
Workpiece tolerances 99
Workpiece whipping tendency 82

X-axis, machine tool reference point 11-13

Y-axis, machine tool reference point 11-13

Z-axis, machine tool reference point 11-131

Printing: Druckerei Zechner, Speyer
Binding: Buchbinderei Schäffer, Grünstadt